# My Least Favourite Species

Tim Biddiscombe

# DEDICATION

For my Nan.

# CONTENTS

# ACKNOWLEDGMENTS

I'd like to thank everyone whose almighty achievments I've referenced herein, Charles Darwin, Denise Breitburg, James Lovelock, Russ George & Mayer Hillman among many, many others – all of them far smarter than I. I'd like to also thank Micks for the wonderfully visualised front cover design and Maree for editorial guidance, and getting me out of a tight corner I'd accidentally written myself into.

# INTRODUCTION
## THE TIPPING POINT (A BRIEF EXPLANATION)

A glass of red wine sits on the coffee table in front of you.

You look around, bewildered for a moment, then the room snaps into focus and you remember you are at a dinner party. You wonder for a moment when the last dinner party you went to was, wasn't that more of an Eighties thing? But no matter.

Up until this point it's all been plain sailing. The evening has gone well, you've met some new people, reconnected with some old ones, said a few pithy things that have gotten a good laugh and avoided mentioning a decades old £5 debt somebody there still owes you. An album by David Bowie is of course playing in the background.

In short, a good evening is being had by all.

You eye your wine glass on the coffee table, it's half full (or half empty?) and been refilled an indeterminate number of times already, but that's no reason to not quaff a little more, since things have been going so well so far.

You reach out to pick it up, and accidentally tip it over. There are any number of possible causes; you were a little distracted and applied too much pressure, someone bumped your arm as you reached out, the few wines too many you've already drank messed with your coordination… but whatever the reason, it's now happened. The wine is no longer happily sloshing around inside your glass. You hope your host is in a forgiving mood, after all they've probably done far worse to this carpet, but that's by the by.

The sudden, violent emptying of the glass (outside of ones' mouth where it belongs) moves the wine glass from one stable state (standing up and full of delicious Rioja), to another stable state (laying on its side and sadly empty). In strictly emotionless, cosmic terms the two states are equally valid (wine in glass, wine not in glass), with only some minor local implications for the state change. Let's look at those local implications.

Your first, irrational thought is, 'Can I reverse the glass to its previous state?'. Not so simple, though. You know that it's possible, even probable to right the glass by closing one eye and concentrating intently. But what about the contents? That doesn't seem nearly so likely - that red wine is seeping into your hosts shag pile carpet even as we speak. Out of the corner of your eye, you note that your host is looking on in horror.

It's more or less at this point that the unforeseen knock-on effects start to occur.

It turns out the shag pile carpet was antique; its brilliant white colour is now stained a spreading crimson bloody hue. Worse still, the wine seeps through and soaks an electrical cable run underneath that is needed for the surround sound system, that's maybe a little worn from all the foot traffic – your host has a lot of parties after all.

You stare at it, paralysed by a perfect storm of surprise, indecision and fuzzy wine head. Remember you're not alone though – how does everyone else react?

Lisa is the nearest, and the first to notice – she dashes forward with hand outstretched to try to catch the glass before it completes its messy state change, alas despite her well-meaning action she is too late.

Bob conscientiously runs into the kitchen to grab a cloth,

determined to in some way mitigate the damage while heading in the opposite direction.

Alistair laughs involuntarily, until he catches the host glaring at him and shuts up.

Everyone else at the dinner party, the majority of attendees in fact simply do nothing. It doesn't affect them, why should they care? They glance at the kerfuffle, then look away, continuing their chatting and drinking.

Beneath their feet however, things are quietly getting worse.

The soaking of the electrical surround sound cables currently passing Space Oddity along their length causes a sudden spark, then a lick of flame. The beautiful shag-pile rug is unfortunately an incendiary bomb in disguise and its lovely dry fibres, so soft on bare feet, catch alight.

Alistair steps back hurriedly from this rapidly evolving drama. Too quickly, his glass of over proof rum sloshes, splashes and literally adds more fuel to what until this point had only the humble beginnings of a fire. Not anymore.

Think fast, hot shot. What do you do?

You can't do anything, is the sad, brutally honest answer.

You just sit there looking stupid with a 'how did this all happen?' expression adorning your otherwise attractive face. There is a small but robust fire blossoming from the rug in front of you. Bob finally dashes in from the kitchen, but his damp cloth just isn't going to cut it.

Lisa steps back also, woefully unequipped to stamp out a fire in those impeccable high heels she picked out for this evening. In front of you on the coffee table, the indifferent, now empty wine glass cares not for its distraught human imbibers, whom until recently had not a care in the world.

It's pretty much at this point that you excuse yourself to go to the toilet and consider doing a runner before the host calls the fire brigade, starts talking about replacement carpet bills and asking if anyone knows a good electrician...

So, what just happened?

I hope there was a point, I hear you wondering from here. I assure you there was... we've just explored a scenario crafted to introduce you to the concept of the Tipping Point. The Tipping Point indeed is the theorised concept of a point in time where the worlds' climate moves from one steady state into another one. As such, it's a somewhat arbitrary and vague idea, the somewhat implausible notion that if we had enough data, we could pinpoint an exact time, to the second when the state changed. Nonetheless, that is what I intend to attempt, in 16 chapters or less.

The nature of the change is simple enough though, the planets ecosphere moves from what we might call a stable state, or at least, a state concordant with our specifically human societal and survival requirements, to a new state, still entirely stable in its own right, but significantly less amenable to us as a species.

That's right, I'm talking about us again, the imbecilic bipeds who still act as if it's a good idea to regularly consume significantly more planetary resources than can be replenished. The shambolic evolutionary offshoots who are still busy coming up with new ways of killing each other over imagined slights and centuries old grudges instead of efforts to bring in resources from outside the gravity well that imprisons us. Nevertheless, the Tipping Point remains true, even if we lack the tools and data to measure it at that infinitesimally fine scale.

The reason it's called a Tipping Point of course, is that there is no reversal possible, no going back to the previous state once it's happened. In our analogy, there's no putting the wine back into the glass once it's unfortunately spilled all over the place.

You can't even go back to the kitchen to get some more wine, as the sofa represents Earth, which is a closed system with no more wine coming into it. The real-world resupply equivalent of the wine on the kitchen side, is likely an icy ball of rock floating out in the asteroid belt, approximately 400 million kilometres away, and weighing billions upon billions of tons.

So near and yet so very, very far.

The accelerating wave of secondary catastrophes that follow a Tipping Point are referred to as 'The Cascade'. The spilling wine, electrical fires and whatnot represent ice ages, wildfires, superstorms, droughts and rising sea levels. David Bowie represents what he always did, peace, creativity, love, and lengthy experimentation with class A drugs.

It's no wonder at that point in the evening that you start looking for the toilet to hide in or the front door to make your getaway. To stretch the analogy all the way to breaking point, the toilet represents a self-contained, underground bunker away from all our struggles, an even smaller closed loop with no hope of escape. The front door likely leads to a small, self-sufficient colony on the surface of Mars, most likely in Valles Marineris, the Grand Canyon of Mars where temperatures can soar up to almost zero degrees centigrade. The colony will, if things go to plan (and when frankly do they not?) be populated by a million close chums of our own Mr Musk, whose extraordinary vision and achievements (despite his outspoken gobby nature on social media) we shall doubtless touch on again down the line.

Perhaps there is yet another option for us to consider in the analogy. Let us for a moment assume you are an honourable person who realises he/she has done wrong and wishes to put it right. Perhaps you won't take no for an answer. Can you really fix it? Do you have the capability? Not by yourself, of course not – no single person does. But what if you joined forces with all the other guests at the dinner party you might yet rescue the rug, get the fire under control and go downstairs to make amends with a doubtlessly irate wine-soaked downstairs neighbour.

Sadly, the rest of the dinner party just isn't interested in helping you, in being on your side, no matter how noble your cause. They say the right things, mumble the right supportive epithets but continue to go about their own business, doing nothing to progress towards a commonly beneficial goal, pursuing their own selfish, insular interests in filling their own glasses still further. They look the other way; it isn't anything to do with them after all. They are completely and utterly wrong of course, but there's no way to convince them of that save the situation getting so bad that everyone including them is personally affected, by which time of course it's far too late.

Whichever of these options you chose, one thing is abundantly clear: the dinner party is over. And it likely took place decades ago, which explains the shag pile carpet if nothing else.

Phew. That's the analogy done with, I hope it made sense, or at least was faintly amusing while it didn't make sense. It felt like it was getting away from me for a while there, but upon re-examination I think I still came out on top.

Back here in reality, the real Tipping Point of our ecosystem was a complex overlapping combination of events with no single precise causative factor, unless you feel like pointing the finger at a certain dominant species with an unhealthy obsession for Pokémon Go. Luckily, we have room ahead of us for a good few chapters to cover these factors in some sardonic detail.

## Part 1 – My Least Favourite Species

We cannot despair of humanity, since we ourselves
are human beings.
Albert Einstein

# CHAPTER 1 - A BEGINNING

Let me take you back 13.8 billion years or so. The neighbourhood looked very different back then, let me tell you. In fact, there wasn't a neighbourhood. There wasn't an anything. It was a much simpler time, in every way conceivable. The universe, such as it could be defined, encompassed an infinitely small, infinitely dense and infinitely hot region embedded in the middle of blankness. Outside that region was nothing. No, I don't mean empty space, not a vacuum. I mean something so, well, profoundly nothing, that physics and mathematics can't begin to grasp at its nature. Not even time existed, that's how far off the reservation we are.

While nothing is known about what happened before this point, there is some interesting speculation that preceding the existence of this tiny spot of something would have been a Singularity of some kind. In case you're wondering, a Singularity is a location in spacetime where infinity exists, so named for their singularly fascinating nature. Fine, makes sense so far – we're already talking about infinity, so a theoretical object that fits those criteria is at least a promising start to understanding what the hell is going on.

Little is known about Singularities outside theoretical models, they tend to make physicists at best very uncomfortable to dwell on, and at worst insane if they spend too long dwelling on them. Since the only place we know that Singularities exist is at the very centre of black holes, it gives rise to an interesting theory that our universe was born from a black hole, perhaps one located in another universe entirely. Legendary sci-fi author Stephen Baxter loved this theory, and once described black holes as 'the place where universes kiss', and if our universe was the child-product of such a union, well, that would be very fitting, if a little romantically inaccurate.

The many-worlds interpretation of quantum mechanics allows for a multiplicity of universes (or 'multiverse' to use its trendier name) where this would be entirely possible, hence this could conceivably be the reason our own universe came into being. Having said that, I'm a curious old chap as you may already have noticed, and I'd want to follow the trail back to what created the universe that created ours, and what created the one before. Down and down, there must have ultimately been a first, primary universe that all others sprang from which is at present sorely lacking an origin story, deity driven or otherwise.

Turtles all the way down, as the old lady once said.

Or perhaps we'd arrive at the bottom universe, only to discover the origin of our ultimate parent-universe is the one at the very top, rendering the procession of universes cyclical and never-ending, twirling through eternity like a strand of DNA. I'd like that a lot, symmetry appeals to me and it would imply an unnatural, even intelligent design to proceedings which would amuse me no end.

Back in our own personal spacetime, for a reason that nobody understands, that tiny pinprick of infinity that was our entire nascent universe suddenly, and quickly started expanding outwards at quite some pace. Explosively fast, some might say.

Ladies and gentlemen, meet the Big Bang. The Big Bang, meet ladies and gentlemen. I'm sure you'll get along famously. The universe is after all a fascinating dinner guest.

The universe surged outwards and physical laws began to take hold in this new and unfamiliar realm. Infinite density and temperature suddenly dropped to finite, and if there's one thing I know about finite, it's that it's a number less than infinity, mathematically speaking. I'm smart like that.

This period of aggressive expansion brought about some rather intense cooling. Just a few minutes after the Bang had happened, the temperature had already significantly reduced down to a balmy billion degrees centigrade, so hats were definitely needed if you wanted to pop outside. The protons and neutrons that made up the hot soupy universe at that point started to combine, forming hydrogen nuclei croutons (I decided to run with the soup metaphor, sorry!). Hydrogen is the simplest atom in the universe, consisting of one proton and one neutron, and even now after so many billions of years have passed, still makes up more than 75 percent of matter known to exist throughout the universe. That being the case, it's no surprise it was first out of the elemental starting gate.

Over the next few hundred thousand years, the atoms started to combine into other, less inert (and more interesting) forms, and physical matter finally started to form. I happen to like matter a lot, most of my favourite things to do involve matter of some kind. Slight variations in the density of the matter allowed gravitational force to do its thing, and slowly, painfully, the objects that would make up the universe as we know it began to form. Clumps clumped together with other clumps, for billions of years. Rinse and repeat. A universe of clumpiness. You can bet that if Stephen Hawking had named his famous bestseller 'A Universe of Clumpiness' it wouldn't have flown off the shelves to sit gathering dust under the coffee tables of the middle classes around the world.

About four and a half billion years ago, when the universe was barely out of adolescence, or at least was at an age where it smoked weed and rebelled against its parents, came the time when Earth formed. Not out of nothingness though, as has occasionally been fatuously claimed – that's after all an absurd idea; the only thing you can create out of nothing is nothing. Even the universe when the Big Bang happened was something, we just lack the scientific vocabulary to adequately describe it. No, Earth formed out of stuff.

Somethingness, if you prefer a slightly more technical term.

Scientists are also pretty sure the origin of everything was all those billions of years ago, and not six thousand years as some have claimed over the years. I acknowledge that may come as something of a surprise to the 84 percent of the human species who claim some affiliation with a deity who traditionally receives the credit for such things. On a personal level, it still absolutely astonishes me that the number of believers is quite so high still, and as an interesting aside, the majority of the unaffiliated 16 percent are apparently based in China where more than 1,000 gods, deities and spirits are regularly worshipped. I guess the Chinese gods weren't so big on wrath and guilt as certain other deities I could name. In any case, I promise this disparity is not why we're here and won't be drummed home at great and borderline insulting depth (looking at you Dawkins) so, let us proceed onwards.

I would like to spare a word or two though for intelligent design, the ambitious, if rather conceited idea that there are aspects of the universe that are more likely to be explained by an intelligent hand, rather than natural, or random origin. While numerous supposedly intelligent beings have argued various aspects of intelligent design, claiming DNA is too complex to form by chance (despite there being billions of years with which to do so), this theory never fails to elicit a smile from me.

What a joke. How, on the Flying Spaghetti Monsters' (now there is an origin story!) green Earth could this epically disastrous occupancy of a planet be perpetuated by a creator. At the very least they would have to be a sadistic anarchist to wish this on a species they had sired, even though they could say it was all our own fault. It is a poor tool who blames his workman, after all. Some of the best advice I ever read, and helps keep me calm in such situations, is that 'You cannot reason someone out of something they didn't reason themselves into'. I don't know about you, but that helps me sleep at night.

My favourite origin story though, has to be from the Norse Mythology. In the beginning, there was nothing. Okay, with you so far, familiar territory and all that. That nothing went by the quaint name

Ginungagap, and some way north of Ginungagap was a place called Niflheim. How an empty nothingness came by a name is rather curious, but such geological oddities could go a long way to explaining some of the Uber journeys I've been on. Niflheim was known as the Mist Home, and from it sprang 12 rivers, which froze as they flowed north. There's a lot of north in this story, but that's where the word Norse literally comes from, so I can't fault it there.

From the ice the first entity to walk Earth formed, a frost giant named Ymir, who like the rest of us was not at his best when he'd just woken up. From the ice then came a cow somehow, named Audhumla, who licked salt in the ice for three days and nights, and from that ice came the first man and first god, Buri. Buri found himself a wife from somewhere, Bestla, and together they had a child, Borr. Then things get a little weird, as if they weren't already. Borr had three children of his own, with Bestla. Clearly things in the Buri household were very relaxed indeed. These three children were called Vili, Ve and Odin.

In contravention to the apparent tradition that was springing up in the Borr household, the brothers found their own wives instead of dating their mother/grandmother. They also went to war with Ymir. In slaying him, the world as we knew it was born. Ymir's blood became the oceans, his flesh the land, his bones the mountains, and so on. Not a bit of him was wasted, which was terribly economical and great to see that even in those early formative days, the Scandinavians were all about recycling.

That's just one theory of course. I could talk about how the god Izanagi and goddess Izanami formed the Japanese islands from Izanagi's dried semen, or how the evil alien Xenu killed a few trillion of his people in Earths volcanoes 75 million years ago, and whose spirits have lived on to plague the living; and only the scientologists can save us. Or I could tell you about Cipactli, the half crocodile, half fish monster from the Aztec creation myth, who was pulled apart by jealous gods, and whose middle section later became Earth. All interesting, and all equally, impossibly implausible.

But let us get real, none of those have any bearing whatsoever on the formation of Earth.

This Earth of which I speak formed from what is known to be a protoplanetary accretion disk, which in layman's terms is a thin circular area of dense gas and dust about sixteen billion kilometres across. At the centre of this disk lies a newly minted star, the accretion disk is in fact left over material from said stars formation.

The cloud of gas and dust is known as a nebula by popular parlance or a stellar nursery if you want to romanticise it. The nebula collapsed under its own gravity, which caused it to spin faster and faster, the larger the knot of matter at the core, the faster it could spin. The great cloud flattened out into a disk, the knot collapsed further, until the pressure and temperature at the core caused a fusion reaction.

A star is born, without the need for any musicals on the subject, though now I'm wondering if I should write one about the solar system. I bet there's some catchy songs about Pluto being a wandering rogue, and Saturn being so beautiful it breaks your heart. In your face, Holst.

Our star, itself barely out of diapers at this point was already going by the cool moniker 'The Sun', or in astronomical terms, just 'Sun', which I prefer out of informal contrariness. All the other planets (both gaseous and solid), planetoids, asteroids, moons and a great many of the comets in our solar system have the same origin story as Earth, which goes a little like this.

This disk of material rotated around Sun, the rotation was caused by the physical laws of gravity and its effect on the matter of the universe, and the act of gravitational rotation caused clumps of matter to form, some larger than others. Once the clumps formed, more gravity happened, drawing them into larger clumps, and from there time and physics did the rest. There is some conjecture that a nearby supernova could have initiated the disks rotation; this is unproven, but the beautiful 'circle of life' symmetry is achingly plausible, even if you're not an Elton John fan. Whether or not that happens to be the case, regular supernovas are essential for new species formation, as their intense explosions scatter valuable heavy elements throughout the nearby stellar neighbourhood. In fact, many of the atoms inside your body were created in previous generations of stars, their supernovae endings seeding our accretion disk with key elements like carbon, nitrogen and oxygen. "We're made of star stuff", as Carl Sagan once famously said, and as always with Mr Sagan, he was not wrong.

Time and physics make the world go around, not money, not love, or even coffee as one board outside a coffee shop facetiously claimed to me the other day. And speaking of time and physics, I'm reminded of all the times when, as a boy, my mum would ask innocently who was doing the drying up, waiting for me or one of my siblings to groan and rise from the sofa to miss some distinctly mediocre 80's television while drying the dishes. While at the time I lacked the vocabulary, the fact that dishes dried perfectly well when left to their own devices was something even my younger self with his terrible attention span was able to notice.

So, if I could get it off my chest, once and for all, for the sake of that poor frustrated boy who often missed the second half of the A-Team. Who's doing the drying up mum? Entropy, that's who. The second law of thermodynamics, that's who. All that drying up was probably character building though, and a well-reasoned scientific debate back then would have almost certainly earned me a clip round the ear.

In any case, however it got started, the disk was comprised from the remnants of the formation of Sun itself, which selfishly took up the majority of the 'clumpiness' of all that floating gas and dust for itself in the very early days. This was a good thing though, if the weight of the 'Sun Clump' as we'll call it had been just a shade under one hundred and fifty thousand million, billion giga-tons it would have failed to ignite its fusion reaction to become a star. Had that been the case, our solar system may well still have formed protoplanets, but the central object in our cosmic tapestry would be what's moodily called a Brown Dwarf Substellar Object (super catchy I know). Without all the blazing solar energy that goes with a living, or at least operational star, life would almost certainly never have come oozing along out of the primordial soup on our lump of rock.

Think about that next time you're considering going on a diet.

It was somewhere around this time that Moon was formed. Oh, how sweet I'm sure you're thinking nostalgically, our ever-faithful companion up there in the night sky formed the same time our planet did. Not exactly, the commonly accepted truth, according to most planetary geologists is much more violent.

While Earth was busy forming, it was to say the least a little distracted. During this period a slightly smaller planetoid (around the size of Mars so it goes) was able to sneak up on it, and clobber it. Earth, which was busily engaged in the act of cooling itself down was instantly molten again after an impact so massive it almost shattered completely. While this interloper was rifling through Earth's wallet for hard currency and credit cards, the impact of this planetesimal threw up an enormous amount of material into orbit. So much so, in fact that the merry cosmic dance that Earth had recently gone through now took place again in miniature above what would one day be our heads. The mass of recently molten material orbiting Earth now itself clumped together under the force of gravity into larger and larger forms, until only one remained.

M-O-O-N. That spells Moon.

How not on Earth could we know this with any certainty? It turns out that all those Moon landings, long suspected of being faked by the tinfoil hat club gave us the evidence for this. The rocks brought back from Moon show that the surface was once molten, as was our own. Furthermore, on deeper examination they show to contain exactly the same proportions of stable isotopes as our own rocks, which could only happen if they had a common origin story.

Moon is more than just a loyal companion to Earth; it is Earths daughter.

While researching this section I was curious to understand where Earth gets its name from. Jupiter was king of the gods after all, Venus the goddess of love, Mars the god of war. Earth? Which god or goddess did we deign to name our cradle of life after? As it turns out, none of them. Our planet is named from the Anglo-Germanic word 'eorthe', which literally means Ground.

Ground???

Honestly, I know primitive man was lacking in imagination, but seriously, we could still somehow dream up the glittering pantheons of Greek and Roman gods with which to adorn our planetary cousins to great aesthetic effect. Perhaps it's because there wasn't a god of thoughtless behaviour, or a goddess of lack of foresight to symbolise our deep and abiding flaws. That may have to be my only consolation.

However, as a friendly gesture, I'd like to stick a name into the ring for future generations of cynics to ponder over.

Geb.

Geb was the Egyptian God of Land, who went around with a viper wrapped around his head and was a member of the Ennead of Heliopolis, which from what I can discern was an Ancient Egyptian version of the Justice League. As a word, Geb is shorter than Earth, you can say it quicker, which would save a lot of time for doing other things. I did consider Horus as more well known, but he's God of the Sky, which would be more than a little misleading. As if accuracy was top of the agenda when pairing up names of ancient gods with specks of light you can see down the end of a telescope that somehow seemed to move a bit quicker than other equally inscrutable specks of light...

Speaking of Jupiter, search further out still in our solar system and you come upon the gas giants and ice giants. The vast, elegant denizens of the outer solar system spin in the darkness, watching over us in silent vigil. That sentence was a very pleasing combination of factual accuracy and waxing lyrical, I should try to do that more often.

These outermost, giant planets are also known as the Shepherd Planets. Since the days of the primordial solar system, with so many balls of rock and ice still flying about even after the formation of the main planets and moons, the early solar system was an extremely dangerous place. The immense gravitational pull of the giant planets has acted as a protector of the inner solar system by 'sweeping' the inner solar system clear of these hazards, pulling them into their orbits and down to a fiery fate, becoming new moons, or even flinging them out of the solar system entirely. You can think of the inner solar system as a hip, happening nightclub, and Jupiter as the surly inscrutable bouncer who is sick and tired of drunk young asteroids trying to get past him towards the bright lights and thumping music.

While we're in the particular neighbourhood of the outer solar system, I have one further thing to say: poor Pluto. Planet for a day. Well, 76 years, 5 months and 29 days, which is a blink of an eye in cosmic terms.

Pluto was discovered in 1930, and for most of the twentieth century enjoyed a status of 'The Ninth Planet', accorded all the respect and privileges that membership of this exclusive club conferred. That all changed however in the nineties and noughties, with the discovery of other objects of a similar size, meaning perhaps it wasn't so special after all. In 2005 Eris was discovered, significantly further out, but was actually 27 percent larger than Pluto. For that reason, in 2006 the category of planet was subtly redefined at the lower end of the scale, and Pluto was unceremoniously demoted down to Dwarf Planet.

Incidentally, all this scientific detail has me thinking that if a deity was behind the creation of all of this, their attention to detail is simply astonishing, and were that the case, and if I was the kind of person who looked good in a hat, I might consider tipping it in their general direction...

'Exactly!', a person of faith might exclaim, but I don't recall any origin parables along the lines of 'And lo, the Great Maker checked the weight of Sun on his or her cosmic scales, and found it was over the threshold to form a G-Type main-sequence star with a Goldilocks Zone ranging from 107 million to 185 million kilometres. And it was good'. More about intelligent design and it's deeply comic flaws a little later on.

I've always liked the term 'Goldilocks Zone'. If you haven't come across it before, it's the informal scientific name for the definition of orbital range around a star where liquid water can exist, known to be one of the key building blocks of existence. Too close to the star, and the water heats into gaseous form, too far out and it solidifies into ice.

It doesn't take an astrophysicist to turn on a tap and conclude that Earth is in one such Goldilocks zone. At 149 million kilometres from Sun as we are, we are scientifically 'just right'. No selfish little girls with adorable blond tresses were harmed in the making of this metaphor, though it did make me want to have some porridge. Back in a mo.

Moving on a billion or so years, Earth, a little older now, but still young enough to have that new planet smell, had cooled significantly, and a steady onslaught of comet and asteroid impacts in those early days had brought huge amounts of water with them to the planet's surface. Those days of endless impacts were long gone now, most of the remaining lumpy matter in the primordial solar system had either already hit something or spun onwards and outwards from the central solar system into what we now call the Oort Cloud. Those that struck left behind copious amounts of that deliciously imbibable combination of hydrogen and oxygen atoms we like to call water.

Above, below and on the surface of Earth, this water of which I speak had changed states back and forth from gas to liquid, liquid to ice and back again thousands of times over as the planet heated and cooled with each impact or major tectonic and volcanic event. Finally, the majority of the water settled down into its comfort zone, the liquid form. Liquid water, you may be surprised to find out, is essential for life to come about, at least, the kind of life we can easily comprehend, being the horrendously blinkered and egocentric species that we are. Our kind of life, incidentally, is known as carbon-based since it makes up approximately 50 percent of the biomass on the planet, and a whopping 18.5 percent of the human body derives from this element alone.

Life.

There, I've said it, the word I've been meandering towards for the past couple of thousand or so words. Life is of course the umbrella term for objects of an organic nature, which possess the capacity for birth, growth, reproduction, occasionally some kind of functional activity in the middle somewhere and then death. But how did it come about on our shiny new marble, spinning along without a care in the wor… itself?

The answer is to be found deep in the oceans of course, and a very, very long time ago. And here we meander up to the point of this opening ramble, something even an excellent person like yourself with your saint-like patience must have started to wonder about by now.

# CHAPTER 2 - LIFE

Having hurriedly covered off a few small topics like the universe and the formation of our solar system, I'd like to make things a little more relatable for you. If I was a Vegas singer, I'd be undoing my bow tie right about now and sitting on the edge of the stage – or as a lazy poseur friend of mine used to do, pocketing a clip-on and strategically hanging a real bow tie around his neck.

I'm going to introduce you to your ancestor. I hope you are presentable, if you are sat around in your pants at least go find a T-shirt and comb your hair. Your ancestor, mine, and every living thing on this planet in fact. The trees and the bees, the bears and the hares, the snails and the whales. Everything, whether it rhymes or not, even Simon Cowell (who most definitely does not rhyme). I expect a proportion of readers will roll their eyes at this point and say something weary like "not more family…", well, I'm afraid so, but worry not they won't be taking up a space at the dining table come Christmas and drinking all your Baileys.

Many moons ago, well, approximately forty-two billion moons ago in fact though I'm certain nobody but my good self bothers to work such things out, our ancestor came into being. A billion years and change had passed now since the formation of Earth. Earth had cooled considerably in this time, a reasonably impressive amount of water (326 million trillion gallons give or take) had condensed into liquid form and covered much of the surface. An atmosphere of sorts existed, albeit filled with hydrogen sulphide, methane, and other nasty toxins that could take your face clean off.

Young Earth was just coming out of the Hadean geological era, which so named for Hades, god of the Greek underworld. This naming choice was something of an understatement, as early Earth was not a pleasant place to live, with regular meteorite bombardment from space, high temperatures and widespread volcanic activity. In many ways it was a lot like Detroit in the summertime. Part of this era also goes by the charming name, the 'Late Heavy Bombardment' (LHB), or 'lunar cataclysm', so known because a number of large impact craters on Moon happened at the same time approximately 3.8 to 4.1 billion years back. This would have had a significant 'impact' on us, because if Moon got a few planetesimals in the face, it's highly likely that Earth, along with Mercury, Venus and Mars would have been subject to much the same treatment.

No-one is quite sure what caused the LHB, but computer models suggest an astonishing ballet of activity in the solar systems early days. A recent theory beautifully called 'The Grand Tack' amazingly has the colossal planets Jupiter and Saturn migrating into the inner solar system for a time, coming all the way in almost to Earth's orbit before 'tacking' back out again, like a cosmic sailboat riding the solar winds. This would have caused chaos as they threw their gravitational weight around, causing a great number of smaller chunks of material from the Asteroid / Kuiper belts into wild and unpredictable new orbits, many of which then would have then intersected with Moon and Earth.

This theory has the beauty of also answering the question of 'where did all the water come from', as the proto-planetoids in question would have been made up of a significant amount of water-ice. A parallel model attempts to demonstrate that Uranus and Neptune switched places in the outer solar system before their orbits stabilised, which kicked loads of material into the inner solar system, and increasingly scientists are coming to the belief that in actual fact both theories are probably correct, and do not remotely invalidate each other.

It's quite the mental picture isn't it, Jupiter at 1.5 astronomical units (1AU = distance from Earth to the Sun) in our night sky would have likely looked a touch smaller than moon does in our sky today, it's something I would have paid folding money to see. The king of the Roman gods, venturing into the solar system to look down upon young Earth, before moving out to take up its permanent vigil.

Once this period of bombardment of poor Earth was over, temperatures normalised, the crust solidified, water vapour settled back down into being oceans once more, and the good times were free to begin. For the purposes of the next section, although gender (and with it gender discrimination) are still billions of years away from being a thing, I'm going to go with 'She' just because I can; I hold the metaphorical microphone after all. I won't give her a name however; my anthropomorphising goes only so far (I now discover), and as anecdotes go its bittersweet at best.

She was a single celled lifeform. One solitary cell. Almost infinitesimally small, she most likely came into existence near a hydrothermal vent deep in the oceans, where dark rents in the seabed spewed up complex volatile compounds along with not a small amount of heat energy from deep within Earth. Among the compounds were nucleic acids and a particular naturally occurring carbohydrate known as a ribose. The compounds were to warm and buffet the water around the vent in some very interesting and unforeseen ways.

Jostled by the violent and random environment, these elemental compounds assembled, disassembled and reassembled themselves over and over into ever more intricate shapes. Then one very particular shape happened. This shape was right. It was good. For the first time in the already long history of our planet, this particular microscopic jigsaw puzzle fit together perfectly, and something so special, unique and interesting happened that I have goosebumps writing about it.

Let there be life.

I want to make a special point of highlighting the randomness involved here, and particularly the long period of time it took for this combination of elements to occur and fuse together into something new. It didn't happen overnight, and it definitely didn't happen at the direction of some creators' design, intelligent or otherwise.

She didn't live very long at all, certainly not long enough to disappoint parents and throw a party while they were away on holiday. In fact, being very honest, she was probably not even the first of her kind to evolve, but during her brief spark of life she achieved one incredibly important thing that her siblings had never managed. Something that had never happened before in all the hundreds of millions of years that Earth had been spinning along. Something that every last living organism on this planet owes their very existence to, even Chad from Nickelback.

She divided.

And these new cells that resulted from the division, her daughters if you will? They divided too. The process of division was something that in a few billion years' time a foolish species of hairless bipeds would come to describe as Binary Fission, which simply meant that both these new cells inherited all the traits of the parent cell. If the daughter cells had had eyes, they definitely would have been their mothers.

On and on the division went, a runaway effect that would forever change the planet, flooding out over aeons throughout the oceans from the origin point. Varieties crept in as natural mutations occurred, and gradually, subtly, over millennia the once silent oceans of Earth became filled with a soupy multitude of single celled life, consuming compounds dissolved into the water such as sulphur, methane and hydrogen. You wouldn't have wanted to go swimming in the sea back then, but then nobody likes being dragged for a swim with your grandmother when you're very little…

In the Nuvvuagittuq Greenstone Belt on the eastern shore of the Hudson Bay, near Quebec, Canada, microfossils have been discovered that date back almost to this period, found in rocks rich with iron and silica that started their lives as hydrothermal vents deep under the oceans. The last resting place of our ultimate ancestors.

A billion or so more years passed, during which time the great party in the ocean continued unabated, while the land stood empty and barren, no doubt wondering whether it was something it had said. No plants existed at this time, just bare rock and dust, an alien landscape everywhere you looked.

Out of nowhere, or at least out of the nascent scientific principles that would one day become Darwinism once they had been given a name, a new variety of single celled life evolved itself into being. This new variety made use of a radical new survival technique, a heat into energy mechanism we now know and love today as photosynthesis, indeed, an offshoot varietal descendent of our common ancestor still uses this technique to this very day.

That's right folks. Instead of burning chemicals dissolved into the water for food, a new generation of the little single celled blighters started burning sunlight, falling freely and endlessly down upon the oceans from above. This new generation of cells took the oceans by storm, since sunlight near the surface was readily available, far more so even than the murky chemical compounds that had already brought them so far.

As chance would have it, a seemingly unimportant by-product of this photosynthesis would turn out to be a gamechanger for our up and coming ecosystem: oxygen. Up to this point the atmosphere was predominantly hydrogen, methane and a little water vapour, but that was about to change in a big way.

The new photosynthetic form of life did not stay single celled for long. Unlike its chemical guzzling ancestors, emboldened and empowered by Sun in a way not seen again until Clark Kent started stepping into telephone boxes, it embraced change wholeheartedly, and the first multi-celled life forms evolved. The term generally agreed amongst scientists to describe the next few hundred million years was 'an explosion of life', and let me tell you, they were not wrong.

Suddenly, life was everywhere, abundant, different, distinct, rampant and rapacious. A vast array of life, now fighting amid the high seas for space and resources, that very conflict forcing yet more change to occur yet more quickly, as per the rules of evolution. This of course was still billions of years before these same rules would be scribbled down just a few meters overhead in a cramped sea cabin by the aforementioned forward (or should that be backward) thinking Victorian gent with a predilection for beagles.

More time passed. A lot of it in fact, like two billion years or so. The waters of Earth heaved with life, growing and evolving, diversifying and digressing. Genetic mutation gave rise to new organic configurations which were tried and discarded, again and again. Those that failed were lost forever, those that succeeded thrived in the next generation.

Eventually, only about five hundred million years ago in fact, the first colonists ventured out of the water onto the land, finding it barren, but with an unspoiled and now oxygen-rich atmosphere ready for them to exploit. Dogged, pathetic even, they persevered, trying and dying by the billions. For hundreds, perhaps thousands of generations many of them returned to the water to breathe, breed, or feed, before finally evolving myriad capabilities for existing on land. They were the forebearers of plant and animal life that we know (and in some cases, are) today. Those creatures that remained in the oceans evolved into their own variants of plant and aquatic life fish, enjoying the same kinds of biological diversity they always had done.

Birth, life, death. Over and over. Endlessly creative, endlessly diverse.

Until that is, a species on the land became dominant. Not the first dominant species mind you, the career-ender of their predecessors came in the form of a meteorite sixty-five million years ago; but certainly the latest, as if that's anything to be proud of.

A species so incomprehensibly arrogant and thoughtless that they wrought changes to their own ecosystem that would now threaten the survival of life in the very oceans that had birthed them all those millions of millennia previously.

A species with so little foresight that they couldn't comprehend that their individual actions would jeopardise their entire species chances for long term survival.

A species that had somehow managed to spread out around the globe, and bizarrely end up stratified in such a way that those individuals with the most greed and callousness coupled with the least empathy controlled all the decision making for the species. Those who could recognise what was happening and knew it was wrong were relegated to the historical side-lines as crackpots and in some cases even labelled terrorists and seditionists.

This dominant species of course, where we come in. And we couldn't wait to start trashing the place.

Up until the 1860s, the conventional thinking (and had been since Aristotle's time) was that life spontaneously generated itself from inorganic decaying matter. Around 2300 years ago, in the third century before Christians are pretty sure something important happened in Bethlehem, Aristotle commented on the apparently common knowledge of the time of a process called 'spontaneous generation'. He cited in his works that fleas appear from putrescence, greenfly from dewdrops, and somehow mice from dirty hay and best of all, crocodiles from rotten logs. These last two seem less common observances and more lazy unsubstantiated thinking to support the conclusion to me, but I wasn't there so I can't comment any more than I already am, suffice to say it's clear why he was a philosopher and not a biologist.

Noted English academic Sir Thomas Browne published a book in 1646 with the catchy title 'Pseudodoxia Epidemica', with the even more catchy subtitle 'Enquiries into Very Many Received Tenets, and Commonly Presumed Truths'. It was one of the first seriously rational critical looks at how we perceive the world around us through the filter of knowledge handed down through the ages, tackling what he described as "false beliefs and vulgar errors". In another time he'd have found himself a job with Snopes or on Mythbusters.

For all its scientific rigour and veracity, his book was not well received. Scottish writer and (pseudo-)scientist Alexander Ross wrote 'To question spontaneous generation is to question reason, sense and experience'. Alexander Ross is not well remembered in scientific circles, and his only major contribution to the human race was as the author of the first complete translation of the Qur'an into English. He faded away into obscurity and finished out his days as a vicar on the Isle of Wight.

Shortly after this in 1668 Francesco Redi, an Italian scientist proved that no maggots appeared in meat if you shooed away the flies, thereby providing strong indicative evidence that spontaneous generation had at best some gaping flaws in it. He shooed away the flies. Literally that was the conclusion of the simple experiment, and may seem unbelievably, blindingly obvious, but in almost 2000 years since Aristotle's time, it didn't appear to have occurred to anyone to try it. For that at least, I take my hat off to young Francesco, a one-eyed king in the land of the blind.

The final nail in the coffin of Spontaneous Generation however didn't come until 1861, when Louis Pasteur conducted a series of experiments to prove that nutrient rich organic material does not develop fungi or bacteria of its own accord, and something extra had to be added before growth would occur. This rather took the spontaneous out of spontaneous generation, and without that the theory finally collapsed in on itself, though I daresay I could find a few people on the internet still convinced of its veracity.

It is somewhen around this time that Charles Darwin strode onto the world stage, and by the time he was done the world would never, ever be the same again.

Charles Robert Darwin was born in 1809 in Shropshire, England, the fifth of sixth children of wealthy society Doctor Robert Darwin. He was also incidentally the grandson of two prominent abolitionists (opponents to slavery), Erasmus Darwin & Josiah Wedgwood. The pedigree was strong with this one, so say the least.

Originally Darwin trained as a doctor, intending to follow in his father's footsteps, and indeed worked with him as an apprentice. Attending the University of Edinburgh, at the time the best medical school in the United Kingdom, he struggled academically, being bored out of his mind in the dry lectures and terrified out of it during the surgical demonstrations. His attention wandering and studies suffering, Darwin fell in with the Plinian Society during his second year, a group of radical disruptive upstarts, free-thinking hipsters born in the wrong era, who delighted in challenging the conventionally held views of modern science. To say this went right up his flagpole would be something of an understatement.

Despairing of Darwin's neglect of his medical studies, his father sent him to Christs College, Cambridge to study a Bachelor of Arts degree in 1828, which was the first stage of qualifying as a country parson, kind of an independent Anglican priest. Clearly at the time, if becoming a doctor wasn't on the cards, the next best thing was the Church. How much say 19-year-old Charles had in this is unclear, but it probably wasn't a great deal. In any case, he spent more time riding and shooting than learning matters relevant to religious theology, no doubt to the increasing exasperation of his father.

It was around this time that Charles Darwin learned of, and fell in with Naturalism, a moderately controversial idea that had been around since the 1750s that only natural forces operate in the universe, contravening the long-held beliefs that spiritual or supernatural forces held sway over us. He indulged in beetle collecting, a popular craze of the 1820s apparently, which incidentally comes across as a slow decade whenever it comes up. Christs College was also where Darwin met John Stevens Henslow, a leading parson-naturalist of the time, which was an interesting role whereby you operate a parish much the same as any other man of the cloth, but operate a belief that god wishes us to understand all of his great creation, and thus to study, collect and categorise everything that crawls, slithers or grows.

This interested me, the concept of a parson-naturalist was an early and admittedly clever attempt to reconcile the ever-increasing pace of our scientific understanding of the world around us with the religious doctrines that had dominated the past and to a significant extent, the present. Darwin himself believed that religion was a 'tribal survival strategy', which I couldn't agree more with, nonetheless he was still convinced that a God still existed that had created everything. But nobody is perfect.

Henslow acted as a mentor to Darwin, and the two became fast friends. Upon graduating in 1831, Henslow recommended Darwin for an extended voyage on the HMS Beagle, an exploratory vessel that was sailing into the history books on a global expedition that included charting the entire coastline of South America. Darwin joined in a self-funded capacity in order to retain control of his findings and thus craft a reputation for himself with the science community of the time.

The trip lasted for five years and Darwin saw many wonders, including the aforementioned charting of the thousands of kilometres of South American coastline, the Galapagos islands, Australia, Mauritius and South Africa before returning to Plymouth. I could write a book just about his journey, but that honour of course belonged to Darwin, and an incredible read it is too. I will give special mention however to the giant tortoise that he collected in 1835, which only passed away in 2006 in Australia at the tender approximate age of about 176 years young, since one assumes it didn't have a birth certificate.

Darwin returned to England, and his geological and natural finds cemented his reputation as a leading biologist and geologist of the time. But he was haunted by all that he had seen on his travels, at the differences and similarities between diverse plant and animal life separated by vast distances. It tormented him for his entire career, driving him to distraction. In 1842 he confessed part of his theories on Natural Selection to botanist John Dalton Hooker, noting wryly 'it is like confessing a murder'. Darwin was well aware of how revolutionary his thinking was, and this held him back from speaking out publicly.

Finally, on 24th November 1859, at the age of 50, he published what would be his magnum opus. 'On the Origin of Species', or 'On the Origin of Species by Means of Natural Selection, or the Preservation of Favoured Races in the Struggle for Life' to give it it's full name (which if you read without once dozing off I salute you) was introduced to the world.

In it, Darwin introduced his scientific theory of Natural Selection, that the diversity of life spanning our planet came about through common descent via a branching pattern of evolution, that ultimately, we can trace back to a single common ancestor. During the endless struggle of life, down through the ages, the fittest survive and pass on their genetic traits to their offspring.

The world at large went into collective apoplexy. Scientists, politicians and theologians alike could talk about nothing else. Karl Marx described the book as a 'bitter satire' and 'a basis in natural science for class struggle in history where Darwin recognizes among beasts and plants his English society'. To this I have to say to Mr Marx, when your only tool is a hammer, every problem looks like a nail. Yet it proved astonishingly difficult to repute, not least because by this time Darwin was such a prominent and respected member of the scientific community.

Reverend Adam Sedgwick, oddly given his title a prominent geologist of the time described the reading as 'more pain than pleasure', but his complaints echoed many of his contemporaries of the time, that of what humanity would become if a non-spiritual origin to mankind was proven, which is hardly a rebuttal. Many of those who spoke out against the work feared we would sink into barbarism and chaos without the guiding hand of a creator to keep us on the straight and narrow, and before you start thinking it, humanity was of course fully capable of sickening and beastly acts even with this guiding light, so do me a favour and lend it no credence. Sedgwick went on to vaguely threaten Darwin with an excommunication of sorts, that unless Darwin accepted the revelation of a one true God then they would not be meeting in heaven. Even John Stevens Henslow, who had known Darwin for 30 years distanced himself, describing the books subject as 'a question past our finding out'.

But it was too late for the naysayers. Pandoras box was open, the truth was out there, and Darwinism was officially a thing. After millennia of implausible flights of fancy about where we might have originated, we finally had an origin story for our species that did not involve mysteriously strict botanical gardens, talking snakes, sinful apples and rib bones.

Going somewhat further back though in the quest for an origin story, brings us to a quaint theory that goes by the name of 'Panspermia'. While that might sound like a science fiction themed porn film, it's actually an amalgamation of 'Pan' meaning 'all', and 'sperma' meaning 'seed'. At least allseed would have sounded less like a film about a guy who has a medical condition that keeps the fruit of his loins eternally young.

This currently unproven theory is that life on Earth did not in fact originate on Earth. It is known that certain types of bacteria can survive the rigours of outer space, for a time at least in some dormant form, awaiting a friendly environment. When that environment comes along, for example a freshly minted planet with an overabundance of water the bacteria can enter an active state and get busy evolving beyond its humble beginnings. The movie Evolution deals with this idea in an intelligent and humorous way.

Some scientists believe that this proto bacteria could be abundant throughout space, existing as part of the interstellar medium in some way, in much the same way as space dust and dark matter. Others believe that the truth is a little less farfetched, at least geographically speaking. They argue that early Mars would actually be the perfect place for life to form, since it has an abundance of certain minerals that would support and even encourage the development of RNA molecules including boron and molybdenum.

Perhaps, just perhaps the primordial soup our ultimate ancestors called home was a bit further away than one might expect and was in fact on the surface of our nearest and dearest, planetarily speaking.

A decent sized meteor impact near to this site on Mars would have flung material into orbit, and some lucky contender microbes could have sailed across the interplanetary space to impact upon another world, a world that one day would develop a species that would think Love Island was the very pinnacle of entertainment.

The theories continue on, becoming wilder and wilder. Directed Panspermia for example has intelligent beings from outer space seeding Earth for their own nefarious, or potentially just mildly curious if insanely patient purposes – the increasingly bizarre Ancient Astronauts TV show deals with this possibility at great length.

Pseudo-Panspermia is the romantic notion that the nebula of gases from which stars and planets form are somehow inherently formatted with life-giving properties, potentially by the universe itself. It holds a certain aesthetically pleasing symmetry to me, despite being almost certainly nonsense.

All of which brings us more of less up to date. In the next section we are going to look at the origin story of nature's supposedly ultimate refinement, or potentially depending on your point of view, where Darwinian evolution went very, very wrong indeed.

# CHAPTER 3 - SPECIES DEVELOPMENT

'Where are you from?', someone asks you at a party. I bet you don't answer: 'The Cradle of Humankind, latitude 25.9254° S, longitude 27.7674° E'.

Maybe you should.

Based about 50 kilometres north west of Johannesburg in South Africa, for those of you who don't read obscure geographical co-ordinate systems, is a very important location in our shared history, and perhaps you're wondering why you've never heard of it.

It's been the subject of intense archaeological (technically paleoanthropological but who can be bothered to type that out) investigation for almost a century now. At these coordinates lies a complex network of naturally occurring limestone caves that a very early ancestor of ours loved hanging out in. It had everything they needed, shelter from the elements… that's about it actually. For everything else it meant a trip outside, or a very long wait until someone invents Deliveroo.

Of particular interest, in a site already full of fascinating finds (more than 1500 at the last count), is Mrs Ples. This is the friendly nickname for a skull found in Sterkfontein Caves at the site in 1947 by palaeontologists Robert Bloom and John T Robinson. It was discovered nearly intact, and by nearly intact I mean it was intact until they accidentally blew it up with sticks of dynamite. And then hit it with a pickaxe. It was definitely an accident though. Fear not, it has since been glued back together and hopefully someone spoke to Robert and John to highlight that Indiana Jones is a strictly fictional account of how an archaeologist should behave.

It remains of some debate as to whether Mrs Ples was actually a female, but what is not up for debate following radiocarbon dating is that the skull is more than 2 million years old, which means we've been around in some recognisable form for at least that long. Sure, we were probably a little bit hairier in those days, notwithstanding the residents of Hipster Shoreditch, but nonetheless disturbingly familiar. Two arms, two legs. Upright. Unable to figure out how to work a VCR. The similarities go on and on.

Upright is an interesting area to explore, bipedalism. This is considered one of the major branches from our shared evolution with the rest of the chimps and monkeys who are our not so distant genetic siblings. Our first bipedal ancestors arose six to seven million years ago and were either one of the Sahelanthropus or Orrorin extinct species that are known to have existed, who both walked the plains of Earth way back then. They had evolved a spine that contained an additional curve which made standing fully upright a stable and not painful position, and shorter arms than legs.

An interesting theory on the selective benefits of standing upright concern the ability to carry things for longer distances, which as someone who was in a relationship for a long time, I can firmly attest to it. It was this improved ability to carry that increased the survival chances of those creatures who liked to hang around in groups, which was probably the origin of our social nature as a species. Since we already know that social networks are the end of our social ability, I rather like the symmetry of knowing where it all started too.

Around twelve years before Bloom and Robinson were detonating fossils in the questionable name of science, Robert Broom discovered the first fossils at the site in 1935. His discovery combined with the potential of so much more to discover in the complex cave system led to a fossil rush. Scientific institutions around the globe hastily despatched their best and brightest, or at least, their youngest and most enthusiastic to explore the site, an endeavour that continues to this day.

Discovery after discovery followed. C.K. Brain discovered the oldest known use of fire, which dated back more than a million years. As recently as 2015 another human relative was unearthed and dusted off, Homo Naledi, who inhabited the area 250 thousand years ago and appeared to share our reverence of the deceased. It was discovered that some sections of the cave system they inhabited were reserved for storage of the deceased, we must hope this was as a sign of respect rather than as a larder.

Do we really hail from this area of Africa then? It's impossible to be absolutely certain this is the precise location, but the diversity of the remains discovered, from across several million years make it a very likely contender. Suffice to say this area was special in our heritage, and we certainly don't have any better prospects to compare it to. All evidence firmly points to humanity evolving in the southern part of what would eventually become known as the African continent.

I'll tell you something else now, which science seems loath to point out in these populist troubled times we are presently tolerating with grim fortitude.

We are all black. At least, we were two million years ago.

Evolutionarily speaking, the selective advantage in darker coloured skin would have been clear, in hindsight at least; protection from ultraviolet radiation and skin cancer. When our ancestors lived in South-Central Africa, blazing, unending sunshine would have been all they knew. Earlier still, when we were covered in hair it wouldn't have mattered so much – chimpanzees have pale skin under all the hair after all, but once we bred out the hair, evolution had to do something to preserve the species.

It's an amazing conceit of our biological wiring, make no mistake while some of the ugliest components of the human psyche may be nurtured by your surroundings and upbringing as an impressionable adolescent, a fundamental uneasiness with people who look different to you is alive and well in the modern human. Back in the day, this trait would have been useful in protecting your tribe against outsiders, what value it in the present day I cannot say.

In the modern world, since we are in all the fundamental ways that count, the same, the very concept of racism is deeply flawed. If little green men would ever get off their collective (assuming they have a hive mind) behinds and visit us, then we might finally have someone to genuinely feel uneasy about, but otherwise if humanity could just stop eyeing each other distrustfully for no apparent reason, the world might be slightly easier to occupy. Taking a macro view as I am wont to do, it's akin to blood cells in the body turning on their compatriots in some terrible wasting disease.

Sociologically speaking, it could be looked upon as the reverse of Dunbars Number, which we will come onto later when we tear tribalism a new one. After all, this innate and subtle uneasiness (easily overcome I might add by travelling and talking) to people who look differently to us that the ignorant and sadistic fan into flames of hatred and racism, may well have had a purpose once upon a time. Our wiring is many layered and difficult to visualise, let alone comprehend. Perhaps we are even seeing some evolutionary response to overpopulation, that in many ways (racism being just one of them) a civilisation is in fact designed from the ground up to tear itself apart once it reaches sufficient mass, in order to preserve the long-term future of the species. I'd like to think there is some kind of plan at work, albeit instinctive, subsurface activity, akin to the vivid dreams we experience before we awake. But as Jules Winnfield points out to us in Pulp Fiction, that shit ain't the truth.

I have to spare a moment to honour the creationists too. Those who, even when faced with an insurmountable wall of data concerning our pre-history over the past few million years, still stubbornly cling to the idea that everything we see around us was spun out of nothingness a short six thousand years ago. There's certainly no excuse for it now, but even a hundred years ago it would have been hard to argue the case. It is a typically, almost hilariously quaint conceit to be able to look at a skeleton of a man next to a skeleton of a gorilla and say to yourself, I see no similarities here. Poor Darwin, he must have felt like he was the only sane man on the planet at one point.

Being able to say the Sterkfontein Caves is where you hail from is factually accurate only up to a point however, as our story is already partway through at this point, and arguably closer to the end than the beginning. As a species we seem determined to 'live fast and die young', just like James Dean never said. It's like stating you are from wherever you were living when you were 21, which for me was above a bank in a dodgy town centre of a dodgy town North of London, and I will never be making a proud claim to that particular heritage when I'm interviewed for a talk show, or for a white-collar crime I thought I could get away with.

All the available evidence indicates that primates emerged as their own distinct evolutionary species around 85 million years back. The earliest found fossils are from around 55 million years ago, though using genetic studies we can look even further back into antiquity. We've all I'm sure seen the exceptional music video 'Right here, right now' performed by Fatboy Slim and directed by Hammer & Tongs (the pseudonym of Garth Jennings and Nick Goldsmith), which depicts a visually stunning, if in places wildly inaccurate account of our evolution as a species from first principles all the way to the present. As inaccurate as it is, it still remains the best widely known visualisation of evolution for public consumption.

Out of Africa, as the unrelated film title goes. Our ancestors moved onwards and upwards, most likely because of dwindling creatures to hunt, even at this early time we were already showing classic signs of using and abusing the natural world around us until it could give no more. In those heady days though the solution was simple, keep moving. Nobody at the time could have ever suspected that one day they might run out of lands to move on to.

And move on they did in a big way. In search of new game, they crossed into Europe from Africa, which were linked at this time via a network of land bridges. They would have crossed from what is now Egypt and Ethiopia into Israel and Saudi Arabia. From there our ancestors would have spread northwards into Europe, eastwards into Asia, and southwards into India, ultimately down through Thailand

and Indonesia before ending up in Australia. From Asia, they would have crossed into the Americas via the Bering Land Bridge. It would have taken a long time, tribes would have risen and fallen, split apart and merged together again, taking thousands of years, but slowly, painfully, all the land areas of Earth began to know the footsteps of man (and more importantly woman).

It goes without saying that the colonisation of the world from our standing start looks a lot like a good game of Risk when you plot the ebb and flow out onto a map.

If our time and location in Africa was the Cradle of Humankind, what happened next was the Cradle of Civilisation. The more I write and think about the topic, the less a veneer of respectability and 'oh-what-a-good-idea' the word civilisation conveys, to me at least. From it we get the words civilised and civility, all intended to portray some ideal of properness, which when you consider the brink of ruin we find ourselves at now, seems somehow trite and pointless.

While there absolutely was a single place (and organism) that humanity hails from, it is commonly accepted that there is no single place where civilisation, such as it is, took place. In fact, it is believed to have emerged in many places more or less simultaneously: The Fertile Crescent, Ancient India, Ancient China, Central Andes and Mesoamerica. Since we have to start somewhere, let us examine the Fertile Crescent, as despite all things being equal, some places are considerably more equal than others.

The Fertile Crescent is funnily enough a crescent shaped region in the Middle East spanning significant parts of what are today known as Iraq, Israel, Syria and Egypt. As the name suggests, it was an area rich in sources of fresh water, allowing its inhabitants to flourish, while being bordered on all sides by harsh landscapes, such as mountain ranges and deserts.

While civilisations emergence broadly happened at a similar time in our history, it's commonly accepted that the Fertile Crescent came first. Located as it is (assuming you're good at geography) as the first stop on the way out of Africa, this represents the place that humanity first stopped, look at the horizon and said: 'It looks nice here, and I'm sick of walking'. Many a dining decision since has been based on the exact same principle.

This area is steeped in human history like no other. It is the site of Jericho, believed to be the worlds' first town, first settled more than ten thousand years ago. Still a city today, archaeologists have unearthed evidence of more than twenty successive settlements on the same site spanning this incredible time period, almost all the way back to the retreating of the glaciers at the end of the last ice age. The freely available freshwater courtesy of many springs in and around the city made it a very attractive prospect, the Hebrew Bible, written around 2800 years back describes it as "the city of palm trees".

The natural resources of the fertile region gave pause to our nomadic ancestors, who for the first time considered an alternative to hunting and gathering. Agricultural farming was born. Animals and crops were raised for consumption rather than hunted down by a people forever on the move.

Within the Fertile Crescent, an area in the East known as Mesopotamia (present day Iraq) on the borders of the Euphrates and the Tigris rivers, gave rise to the world's first empires. The first of these was Sumer. From about 7500 years ago, the Sumerians were formed from primitive farmers and fishermen of the time who found that they were stronger together, less vulnerable to the elements, to predators, and ultimately each other. They banded together into larger and larger units until the Sumerian civilisation organically came into being over a period of 1500 years or so, though in reality it was a loose coalition of around thirty city states with common goals of first surviving and then thriving.

Among these city states was Eridu, thought to be the oldest city of them all. It was apparently the home of Enki, Sumerian god of water, knowledge and creation. I still find it strange that human beings appear to possess an innate desire to give away the credit for their achievements to higher powers. Nowadays very little remains above ground of the ancient city, though archaeological expeditions in between the regions trademark turmoil and conflict have yielded fascinating finds of this long-forgotten milestone in our species' history.

Sumer was the birthplace of many things we take for granted today, the wheel, writing, farming and irrigation to name just a few. The oldest known libraries dating back 4500 years emerged in Sumer, representing a pooling of knowledge hitherto undreamt of since the time of the dark recesses of certain limestone cave systems outside Johannesburg. One of the greatest works of Sumerian writing was the 'King List', which was a stone tablet that recorded all the kings of Sumeria (no queens I might add), including facts like reign length and which city the king belonged to.

The curious thing about the King List is the length of tenure of some of the kings, often stretching into the thousands of years, which shows either they were extraordinarily long-lived, or (and this is my personal preference), they were making this shit up as they went along. It's highly likely that such an important document as the King List would be politicised and spun, emphasising those in vogue, and decreasing the importance of those who were not.

Hot on the heels of the Sumerians, relatively speaking came The Akkadian Empire and then the Babylonians. The Akkadians were the first empire to flourish in ancient Mesopotamia and encompassed the former territories of Sumer along with its neighbours, as well as great swathes of land of what is now Saudi Arabia. It formed around 4300 years ago and lasted approximately 180 years, which in those days of short lifespans was nothing short of astonishing. Its collapse is thought variously to be either because of a severe drought, or a curse brought about by raiding the wrong temple causing some of the gods to withdraw their support from Akkad. I know which one my money is on.

With the fall of the Akkadians came the Babylonians, who owe its short but memorable tenure to the great Hammurabi, a progressive leader who among other things instituted a scaled punishments system for transgressing the civilisations rules. While that may not sound like much, the traditional crime and punishment system up to that point generally involved sticking something sharp through the person you thought had wronged you, so proportional response was a huge step in the right direction.

The same scenario played out around the world time and again. Nomadism was gradually replaced by sedentism, where staying in one place for a long time became very fashionable, even before the advent of sofas and Netflix. Animals were domesticated, forests were cleared in favour of farmland, crops and grains were selectively bred, leading to surpluses of food. Take one guess what tends to happen when you have a surplus of food. If you had "an explosive population growth", give yourself a gold star.

Don't get too hung up on the 10k to 12k year timescale either. While this is commonly accepted to be the point of widespread human species migration around the globe, much older artefacts have emerged in the strangest places indicating a very messy and unstructured colonisation of our blue green mudball, just like all other human endeavours. There would have been many false starts, branching offshoots that went nowhere leaving curious relics behind them, our chaotic geographical progression a curiously haphazard echo of our evolution.

For example, the Woman of Willendorf is a small figurine depicting the female form, discovered near Willendorf, Austria in 1908 by archaeologist Joseph Szombathy. Its voluptuous form indicates it was likely to have been used as a fertility symbol, a precursor to Venus, goddess of love. While Venus didn't start plying her trade until about 2300 years back, the Woman of Willendorf has staggeringly been radiocarbon dated at between 25 and 30 thousand years old. It's now on display in the Naturhistorisches Museum in Vienna, if you're in the area I highly recommend a visit to pay homage to our earliest ancestors, who even in those ancient times, just wanted to get it on.

A series of climate changing events may also have unintentionally aided the formation of these early civilisations. The imaginatively titled 8.2 Kiloyear Arid event, which occurred 8200 years back and the 5.9 Kiloyear Interpluvial, dramatically increased the desert areas of the planet as semi-arid areas dried out completely. This increased competition for the available resources and is thought to have triggered many violent skirmishes between groups, who up to that point had coexisted peacefully. Ultimately it led to the creation of the first walled cities as more and more people banded together to pool their energies and increase survival chances.

There is a lot of talk also about the relative 'ages' our species has gone through, starting of course with the Stone Age, which went on for a long time, around 3.4 million years by most estimates. The term Stone Age incidentally denotes the use of stone to make tools, not the use of stone tools as is commonly thought. By contrast, only six thousand years or so have passed since we left the stone age behind, which is approximately 0.09 percent, or one 1166th of the time since our species started walking on two legs around seven million years ago.

Fun fact, I actually own a Neolithic burin, which is a handheld, wedge-shaped stone chisel used by prehistoric man for engraving and carving. Not a replica either, this is the real deal, only several careless owners. I picked it up from an archaeology collector some years ago, and I keep it to remind me of how far we haven't come. It's good to not get ideas above one's station, especially since as a species that is something we are so very, very good at.

The Stone Age gave way to the Bronze Age as I mentioned somewhere around six thousand years ago, when primitive metallurgy and alloying became popular and widespread. As an alloy of copper and tin, bronze was extremely durable especially compared to its component metals and represented a fundamental shift in the sophistication of thinking. This led to a new relatively post-scarcity era of prosperity, where the number one thought on everyone's minds wasn't how am I going to survive until tomorrow and what shall I eat, gave rise to abstract thinking and creativity, a bandwidth that our species had never before been afforded, certainly not at scale.

Finally, according to the three-age model of prehistory, humanity slipped (kicking and screaming no doubt) into the Iron Age around 3300 years ago, though different civilisations underwent the shift at different times as information slowly disseminated around the globe. Bronze, for all its great durability went under the hammer, and iron took its place, an even stronger metal, all the better to make pointy things to stick each other with. This era lasted until around 1500 years ago, roughly the middle of the fifth century by the measuring of anno domini.

From here the term 'modernity' creeps in to arrogantly name our current epoch in some symbolic way, but should we make it through this particular roadblock our species face, our descendants in ten thousand years will hardly refer to our present era retrospectively as the 'modern time'. Typical egocentric anthropocentrism. Modernity encompasses abstract eras over the past thousand years that you may be familiar with such as the Renaissance, the Ages of Reason and Enlightenment.

Modernity gave way to the Industrial Age, where we found new and innovative ways to damage this precious ecosystem we owe everything to. All of which brings us up to the present, and the so-called Age of Information, whereby our collective species has developed an unhealthy obsession with digital devices, in particular the not-so-smart phones we are all guilty of staring at for upwards of 18 hours per day.

Over the millennia many historians, philosophers and charlatans have attempted to further compartmentalise the ages of man. Hesiod, a Greek Poet who lived about 2700 years back came up with five ages. Golden, Silver, Bronze, Heroic and Iron, with the further back you go, the more gods and humans apparently coexisted happily. You'd also be forgiven for assuming Hesiod was a big comic book fan, but he was attempting to convey the slightly rose-tinted view of history, and the general conviction, even three millennia ago, that things were much better in the past than the present.

Saint Augustine of Hippo had a different view however, when he put forward his six ages of history around 1600 years back. He argued that since Earth was 6,000 years old, each age lasted approximately 1,000 years, with Adam starring in the first age, Noah in the second etc. Under this plan however the sixth age has been running for 2600 years or so without end, which is where the model starts to break down.

Incidentally, Augustine had some reasonably progressive views for a man of his time. He was strongly against slavery, especially that of children. The emperors of the slowly disintegrating Holy Roman Empire had long allowed this practice, allegedly to prevent the parents who couldn't look after a child from murdering it instead. The Child Support Agency and the NSPCA were eons away from opening their doors, and apparently the practice of infanticide was commonplace prior to this child slavery edict. After all, why kill an unwanted kid when you can sell it for some cold hard cash. Humanity went through some pretty inhuman times in the fourth century.

There is so much we don't know, so much lost to the mists of time. Given how little was written down, it's actually amazing we know as much as we do about the past ten thousand years, yet there are enormous gaps, and with each new discovery the history has to be redrawn a little as assumptions are proved right or wrong.

So perhaps when someone asks you 'Where are you from?', perhaps the best answer is, 'It's complicated'.

# CHAPTER 4 - FRAGILITY AND INTERDEPENDENCY

A fundamental arrogance permeates the centuries of man's dominance over this planet. I use the term man very specifically before someone castigates me for a typically male viewpoint – more on that later when we take a baseball bat to our patriarchal society. That arrogance can be expressed very simply: the beautiful blue-green jewel laid at our feet was not put here for our benefit. We happened upon its miraculous resources, took them as our own, heralded their existence as prophetic evidence of our own greatness and twisted them for our bizarre industrial purposes. One fundamental truth remains that many of the world's leaders and titans of industry seem to fail to have grasped – this planet can be destroyed.

That is of course totally inaccurate. Barring something impossibly catastrophic like a black hole forming at a particle accelerator (one of the fun toys scientist like to use to poke and prod the universe) like the Large Hadron Collider (LHC) at CERN in Switzerland, this is actually beyond our species capabilities.

A few fun facts about particle accelerators while we're on the subject. While you may be aware that they are used to accelerate subatomic particles to appreciable fractions of the speed of light for the purposes of spectacularly short-lived, if highly educational collisions, did you know there are more than 30,000 of them in the world? Sounds huge doesn't it. Well, only a few of them are actually 'huge' on the scale of the LHC (with a circumference of almost 27 kilometres), the majority are small and used for specific materials manufacture or medical research. Only the big boys and girls get to look at the big questions like the God Particle (a.k.a. the Higgs Boson) that so captured the medias imagination a few years back.

Another fun fact related to particle accelerators relates to one of my favourite conspiracies of all time. When the LHC was being constructed, it encountered numerous delays due to accidents and unforeseen issues. Nothing surprising there, considering it's one of most complex and highest profile megaprojects our species has undertaken for decades, at least until the ITER fusion reactor comes online (more on that later too).

However, that wasn't enough explanation for distinguished physicists Holger Nielsen and Masao Ninomiya, who publicly postulated that the magnetic resonance issues repeatedly experienced during early testing could be being influenced by the future in some way. Their outlandish theory was that either the God Particle so resisted scrutiny that the Universe itself wouldn't permit the creation of a machine to detect it (which I rather liked), or the implications of detecting it proved so terrible that future humanity devised a way to influence the past to sabotage the LHC to prevent its detection (which I liked even more).

Predictably, their worries were unfounded, and in 2013 the Higgs Bosons existence was confirmed. Though to be fair since then Brexit and the Lunatic in the Whitehouse have both started ripping up society in their own inimitable ways...

Or were they unfounded? The LHC isn't even halfway through its life, with operations and upgrades planned until at least 2035. Perhaps the problem humanity isn't meant to solve, or cataclysm the LHC will bring about hasn't yet been tackled, and the construction of the magnetic rings was the easiest place to try sabotage, the weak link in the chain of events if you will.

Next on the list of problems that the collider will attempt to garner data on is why in the Flying Spaghetti Monsters green Earth is there so much more matter than anti-matter in the universe, and the mundanely titled hierarchy problem. The hierarchy problem is the complaint among particle physicists that the universe doesn't make sense. A plaintive protest to be sure, and more specifically the fundamental forces of the universe, gravitational, electromagnetic, strong nuclear, and weak nuclear interactions are not equivalent. Perhaps some of these noble quandaries would be better off eluding us for the time being, at least until we've curbed a few of our excesses around overpopulation, global warming and climate change.

Remind me one day to write a book about my favourite conspiracy theories, the intellectual stretch is a great way to while away an afternoon, and even beats day-drinking.

Back on the topic of planetary destruction, the late and very great George Carlin once highlighted the inaccuracy of well-meaning individuals who wish to 'save the planet', and the great fallacy this implies. In his own immortal words, he typically cut right to the heart of the matter.

The planet is going to be just fine - it's the people that are fucked.

Sorry for the F bomb, I do try and keep them to a minimum, and in this case, it wasn't even mine, blame George. Of course, many people did. What he was highlighting is that nothing we can do to the planet has any impact over the cosmic timescales that Earth will exist. We are a blip on the radar, a minor blemish, soon eradicated by our own combined inability to keep from consuming all the resources we needed to feed our lavish lifestyle and to keep it in our pants, so to speak.

Plastics in the environment? Not a problem, over a few hundred million years the land is completely renewed via geological subduction. Given Sun has a lifespan in its current main sequence yellow state of another five billion years, it's hardly a blink of an eye in the grand scheme of things.

'Well, what about if we launched all of our nuclear weapons' I pretend I heard you ask, exasperated? There are a shade under 15,000 nuclear weapons in existence today (down from an absurd cold war high of more than 70,000). Less than 4,000 of the ones still around are mounted and flyable, but let's just say for the sake of argument that somehow all 15,000 of them managed to be detonated. That'll sure teach those educated in different ideologies the error of their ways.

Fifteen thousand nuclear explosions. That's more than enough death to go around for the human race, enough to pretty much guarantee the extermination of well over 99 percent of the population, and the ensuing clouds of fallout will ensure the remote living rest aren't likely to be around long enough to rebuild and repopulate. Some privileged (I use the word ironically here) few will survive for some decades in bunkers, until they either go insane or run out of resources. Either way, there's no future underground in this scenario, no preserving of life - our way of life is out the window at this point, whatever it once was.

In this nucular scenario (as those considerably more notorious than me have uttered) of course, we're done. The cockroaches will doubtless one day puzzle over our ruins, wonder what those strange Golden Arches and Green Mermaids on each street corner once meant, and crawl on with their lives, safely ensconced as the new rulers of this planet.

The planet Earth keeps on spinning, caring not a jot for the self-destructive impulses of its supposedly primary inhabitants. Moon looks down indifferently. Sun has no opinion at all, that's how little we mean, how little impact we have.

Despite how small we are, we continue to treat the entire planet as our own personal dumpster fire. Let's take a look at what we're doing to the oceans.

Climate Change is one of the more abstract and indirect ways we hurt the oceans. It takes time, happening so slowly in fact we barely have the tools or patience to measure our own impact over such timescales. We do of course have ocean-hurting weapons at our disposal that are much, much more direct.

Take oil spills for example. Over the past half century industrial accidents have time and again spewed ancient biomass all over our formerly pristine environment. A veritable out with the new, in with the old, ruining entire regional ecosystems, all sacrificed at the altar of Industrialised Capitalism.

I'm sure you will all recall the Exxon Valdez oil spill which took place in 1989, the intense public outrage at one of the worst environmental disasters ever to have been caused by mankind's greed and ineptitude. In real terms, outside of oil spills, the only worse things which have happened on our blemished gem would come under the category of acts of universe (I won't be using the G-word, sorry) like asteroid impacts, solar flares and super-volcanic eruptions.

Through a combination of criminal negligence and abject incompetence, the tanker Valdez ran aground at 12.04am on March 24, striking a reef in the Prince William Sound off Alaska. Over the next 72 hours approximately eleven million gallons of oil poured into the defenceless, unsuspecting oceans. The impact on the region was nothing short of apocalyptic. More than a quarter of a million birds had perished in a matter of months, and the data is unavailable to estimate the total loss of marine life, suffice to say it was utterly devastating, with an ecological impact that is still felt today.

Worse still was the Deepwater Horizon rig incident, which at 7.45pm on 20th April 2010 (my mums birthday if anyone is curious) suffered an uncontrolled pressure blowout, leading to eleven human deaths, and an incalculable effect on the oceans with more than two hundred million gallons of crude oil discharged this time. As is becoming depressingly familiar, the employees blamed the company for cost-cutting and negligence, while the company blamed the employees for incompetence.

Less well known, at least to me was the 'Ixtoc I' spill that occurred in the Gulf of Mexico in 1979 which resulted in yet another environmental catastrophe. Due to the relatively remote location and submersible nature of the rig, the exact spillage numbers are unknown but based on what information is available, it seems likely that tens of millions of gallons of oil were involved.

Two points though are worth highlighting about the Ixtoc spill. Firstly, in an attempt to recover the situation, the surface oil was 'burned off', which would only have moved the pollution problem from marine to atmospheric, and in fact may be the first and only case of brushing the problem over the carpet. Secondly, the oil company involved managed to escape almost all repercussions including compensation claims due to being state owned by Mexico, and as such was able to claim sovereign status. Sorry birds and wildlife, we'd just love to help you out, but we have diplomatic immunity...

There will be more on our unhealthy obsession with hydrocarbons later.

Changing subject to one slightly less greasy, do you recall our evolutionary cousins who stayed in the sea? Well it turned out the vast majority didn't (at least to our knowledge) develop anything like our intelligence, culture or civilisation which, when you look at what we're doing to the planet, even without bringing up Jersey Shore, is hardly saying anything. Based largely on this disparity in intelligence, we've been freely catching and eating the poor submarine inhabitants of our planet for millennia now. And as our population grows, so apparently has our fish and chips habit.

I am of course referring to overfishing. Don't get me wrong, I get it. We've backed ourselves into a culinary corner due to our own overpopulation. Billions of people around the world rely on fish as their only source of protein. I'm not going to sit here and deny them that, but nor am I prepared to endorse it. I didn't ask for those billions of people to occupy the world. For that we have our Wiring (and lack of self-control and self-awareness to overcome it) to blame.

This is nothing new though. Overfishing has been going on for centuries, almost a thousand years by some estimates.

It goes a little something like this. Thousands of years ago, some hunter gatherers got real sick of hunting and gathering. They stopped on the shore of an ocean or large lake and switched up to become farmers, herders and fishermen. While tough at first, they survived, and eventually thrived. Eventually a coastal village prospered on the site, growing into a town, and then into a city. Yay dawn of civilisation etc. To keep up with all that growth more and more food is needed, and what better source is there than that lovely ocean lapping at our toes. Invasive techniques such as bottom trawling, or simply conventional fishing done way too often devastate the population of fish in an area to the point where there aren't enough fish left for it to be worth fishing any more.

Naturally of course, the fishermen realise the error of their ways and took up beet farming.

Nope, only joking, they move further out to fish in a different area and do the same thing all over again, over and over until entire regions lay devastated, as many as we can reach in fact. Despite this, nature is a resilient bitch. Historically, marine population recoveries do take place when the fishermen move on, only for another generation of entrepreneurial humans to exploit them yet again, this time with more invasive and effective techniques. This leaves fewer and fewer fish to recover the population each time, thereby taking longer and longer, and at some point, not enough will remain to undertake the repopulation.

It's just as the old saying went, give a man a fish and he'll eat for a day, teach a man to fish and he will obliterate an entire ecosystem and ruin his own regional, and eventually global economy.

The persistent overfishing globally has continued to the present day. In fact, according to the WWF (the environmentalists rather than the wrestlers), 85 percent of the worlds fishing areas are either depleted or well on their way to becoming that way. To an outside onlooker, say a passing extra-terrestrial, it would look a lot like a war of extermination. 'What did the poor fish do to the humans?', E.T. might muse to himself before waving his tentacles dismissively and moving on to a planet more deserving of their life-saving technologies.

Putting the plight of the fish to one side for a moment, our relentless conquest of the oceans aquatic inhabitants in the name of economic drivers (normally a small group of people making a lot of money putting protein into a large group of peoples' mouths) has not been without it's collateral damage.

Dolphins, believed to be the second most intelligent creatures on Earth (behind us amazingly, no matter what Douglas Adams may have postulated) are routinely killed 'by accident' during our invasive fishing expeditions. It happens so often it's given the appallingly understated name 'bycatch', as if calling it something innocuous would somehow make it okay.

Hundreds of thousands of dolphins fall every year to this disgusting practice and can even end up being processed into tins of tuna in the less reputable canning plants of the world. Ever looked at a can of tuna and seen the words 'Dolphin Friendly'? Incredibly, that doesn't mean no dolphins died in this fishing expedition, it means that the fishermen didn't deliberately target a pod of dolphins feeding on tuna, which is apparently a great indicator of tuna in the area. So, from now on every time you see the phrase 'Dolphin Friendly', remember it means we only killed dolphins by accident and not on purpose. Enjoy your tuna on rye.

Honestly, it's no wonder that there is no serious attempt to translate dolphin clicks and whistles into human language, as no doubt the first words out of their bottle-nosed snouts would be 'You utter bastards', or something even more profane. That's not strictly true of course, in my research of being able to write that snarky line I did find a couple of very promising lines of inquiry on dolphin translating, including some success by a team run by Denise Herzing (weirdly there are a lot of Denise's working in science) operating in the Caribbean. They have been teaching the dolphins simpler, more recognisable clicks and translating them back when the dolphins use them. It's progress of a kind, even if it is forcing them to use a more basic language in order to speak with them. That said, it doesn't feel dissimilar to the memes that the kids use to communicate today, so perhaps it's merely a sign of the times.

Recent research in fact indicates their language may have a sophistication approaching our own, making it all the more appalling that catching them in a net and letting them drown isn't a crime of any kind.

They may be the smartest, which is frankly heart-breaking, but dolphins aren't even the largest seagoing creatures we go out of our way to abuse. For at least the last four thousand and change years we've been steadily hunting our way through the worlds' population of whales with steady increases in brutal efficiency.

You'll have heard of the Inuit people of the Arctic regions of Greenland, Canada and Alaska. Their hunting of whales has, for thousands of years been both ethical and sustainable. They caught what they needed, and literally every part of the creatures was used in some way, from food to toolmaking, from basket weaving to house roofs. The Inuit population was relatively constant, probably in large part to their inhospitable (by our standards) environment, which introduced some natural caps on sensible community sizes and consequently on how many whales they would need to hunt.

If whaling had remained that way everything would have been Hunky Dory (coincidentally the title of my favourite David Bowie album, released 1971), but of course it didn't. As a species, we just can't help ourselves. As ever, economic drivers and capitalist expediency led our species to hunt more and more of the peaceful denizens of the deep.

The 1600s saw huge exploitation of whales indigenous to the coastal waters off America. At one point in North America, whale oil (derived from whale fat, or blubber) was the preferred fuel for lighting lamps, which gives you some inkling of how many whales would need to have been caught for that to have become fashionable. Fast forward a century, and the coastal waters off America had run out of whales for some strange reason... not to worry, the ships were getting bigger and better, we'll just head further out; the Arctic and Antarctic might be a bit chilly, but we'll wrap up warm, and those whales won't know what hit them.

Into the 1800s, and whaling is bigger than ever. Ships are now steam powered rather than sail, giving them far greater range and manoeuvrability, harpoons are now fired from deck-mounted guns rather than thrown, making them far more likely to score a lethal hit. Thousands of ships trawl, literally the oceans of the world, seeking out the ever more elusive whales. I don't use elusive because they had learned to avoid us mind, they had never encountered anything like us before and the slow progress of evolution was going to be no help at all, no, it's just a steadily decreasing population makes them harder to find and kill.

What makes it worse of course is that like the dolphins, the whales communicate with one another. Actually communicate, the free transfer of information back and forth between intelligent entities. You know, the kind of thing a sentient creature who doesn't deserve to be murdered for economic reasons by another species might do.

Sure, the communication is abstract as all hell, but the beautifully complex and rich language of Whalesong, again yet to be translated is sure to contain concepts like 'danger' and 'warning', but probably lacks the subtlety of "I've just been shot by a pointy stick when I came up for some air, how dickish is that?". And in any case when 'danger' is being broadcast in Whalesong from every corner of the oceans, where exactly do you turn?

It must have been a very scary time to have been a whale.

Back to the tale at hand, by the 1900s human beings finally released that something had to change. In 1946 the International Whaling Commission (IWC) was formed. The cynic in me notes that by this time there were cheaper fuels than whale oil, cheaper foods than whale meat which, coupled with the scarcity of whales now (since they were on the brink of extinction) meant that the capitalist opposition didn't care enough anymore to block the IWC formation.

However noble they were in intention, the IWC conventions had little to no effect. The regulations were ineffective, the quota limits given out were ridiculously, unreachably high, and the populations of whale continued to decline. I picture some 1940s tobacco-esque lobbyists distorting the vision to suit their ends, ironic given in those days they were surely smoking their heads off while they bribed with dinners, cars and houses to sufficiently influence policy.

Things eventually changed for the better, in 1971 the USA banned all commercial whaling, and with ridiculous amounts of negotiation that took decades, established whaling-free sanctuaries in the Indian oceans in 1979, and around Antarctica in 1994. Wow, that was like practically last week, I'm sure you also thought; at least if you were as old and cynical as I am/feel.

The whale population slowly started to recover.

Only joking, of course it fucking didn't. Sadly, those sinister cigarette smoking shills back in the forties, like the soulless corporate gangsters that they were, introduced a provision into the original whaling commission to allow for the hunting and killing of whales for 'scientific purposes'.

What a convincing argument it must have been in those days. I picture some earnest, bespectacled and lab-coated advisor, bought and paid for by the whaling lobby. He would have passionately laid out scenarios whereby, what if some sickness was sweeping through the whale population, which could only be treated by capturing and analysing a healthy specimen in order to develop a cure? Well of course, the IWC would have replied, excellent point, lets add an exemption in for scientific whaling.

And finally, our advisor would have pointed out, if we catch a healthy one (for an extremely good reason obviously), we might as well sell off the parts after we've finished our detailed study - waste not, want not, right? The IWC would have looked around at each other for a moment and nodded vigorously. Makes sense, right?

How those shills must have snickered into their expensive drinks in their expensive clubs.

The consequence of the appalling scientific whaling exemption is still felt today. Japan, while not the only ones, are by far the worst offender. As if all the hentai comics and schoolgirl-pantie vending machines weren't bad enough. With more than 90 percent of the World's whaling fleets permanently moored up and backed up by the latest gadgets such as satellite imagery and global positioning, they are free to profitably hunt the remaining whales with nothing and nobody to stop them. All in the name of science, though you won't see anyone with a lab coat anywhere near the corpses of these majestic creatures.

The public perception naturally, with the possible exception of the occasional news article or desperate Greenpeace intervention attempt, is that whaling no longer happens, at least not on a scale to affect the species. Which continues of course to quietly decline.

While on the subject of whales and Whalesong, it would be remiss of me to not digress into a rant (rantgress? digrant?) into the appalling military practice of low and mid frequency SONAR. While apparently handy for insufferably paranoid nations and their respective military industrial complexes on protecting their imaginary watery borders (& annual budgets), they also wreak unimaginable harm on the peaceful inhabitants of the oceans, especially among the whales.

Where to begin. We've been pummelling the oceans with SONAR (SOund Navigation And Ranging) waves since the 1960s, but the frequencies have been getting lower and lower as they get more and more precise, driven by the usual arms race of harder and harder to detect ships and submarines.

Stealth, literally the silent killer.

How is the SONAR bad then? Well, for a long time it was confusing and disorientating to the creatures, which is bad enough when you're a seagoing race being hunted mercilessly by a creature you can barely even perceive. As if that wasn't bad enough, those aforementioned merciless creatures hunting them realised the effect the SONAR had on them and started using it to drive and panic the poor beasties, making them easier to hunt and kill. It wasn't as if the whalers had a big enough advantage already now was it...

That was just the tip of the iceberg though, to mix my ocean-going catastrophes for a moment (and more on the ice later). It turns out, at down around the three kilohertz level the SONAR becomes painful to whales. Deafeningly painful in fact. It drives whales in their droves to beach themselves trying to get away from it. It drives them to race to the surface too quickly, causing a kind of decompression sickness. It causes them permanent hearing damage, and drowns out all other undersea noise, ruining forever the haunting songs that have bounced back and forth between distant pods for tens of thousands of years.

The story of the ocean, sung for generations, taught from adult to child, and lost forever because a certain species of arrogant land bound mammals convinced themselves they were better than the whales.

While utterly tragic, even the whales are the tip of a much greater catastrophe of course, at least from our own selfish standpoint. Like it or not, in fact, agree or not, the fact remains we as a species are poisoning our entire planet, administering a lethal dose of toxins that the unsuspecting inhabitants of Earth seem to be entirely unaware of, despite collectively holding the syringe.

Solutions have been proposed, naturally. Even in our callous, unforgiving culture, there are some people who do care, who do see the big picture. People like ecologist Russ George. "We know that since 1950 we have lost 12 entire Amazons worth of biomass from the oceans", he pointed out matter-of-factly in a recent interview. And he's right, our collective avarice and hubris are well on the way to turning the oceans into deserts.

According to him, the first thing we'd need to do is trap and store again the hundreds of billions of tons of carbon that our relentless pursuit of energy to drive our planetary economy has freed up from the ground to choke us in the air. But can we put the genie back in the bottle that easily? Well funnily enough, Russ has a plan, and it all comes down to iron. Very Tony Stark of him. It might surprise you to know that while a common element on land, iron is the one of the rarest elements in the ocean, and yet is essential for seagoing photosynthesis by the algae and plankton that make up much of the seas' inhabitants.

Russ argues that dumping iron into the oceans would achieve a wonderful domino effect, the plankton blooms, the population goes through the roof. And what does photosynthesis need as a chemical input? You've guessed it, carbon. An explosion of plankton globally would literally suck the carbon out of the atmosphere, turning some of it into oxygen, something we would also find quite handy, and the remainder would be absorbed, to eventually sink to the ocean floor when the plankton naturally expire.

Whether the plan will be acted on however, is another matter altogether. This iron/carbon solution has actually been tried and tested by my new hero Russ. As part of a local ecological effort to restore overfished salmon populations at the behest of the Haidal Gwaii islands, he dumped 120 tonnes of iron sulphides into a 100-kilometre square area of the Pacific Ocean off Canada.

It turns out, amusingly enough though, that Russ George had been around the block a few times. He knew how to get things done, and he apparently also knew what he could not get permission to do. Naturally, being my kind of ecologist, he went ahead anyway. The Canadian government took a dim view of this, and despite counterclaims that they were involved and informed from the very start, Environment Canada raided Russ's offices, and bizarrely deliberately destroyed all his data so that nobody could learn from it.

Thanks EC, you guys are the absolute best. Despite this frankly remarkably stupid decision, the operation is widely considered to be a success, regional salmon populations have since increased by an estimated 400 percent over subsequent generations since his actions.

It's bittersweet ends to great stories like this that are the source of my cynicism that the damage we've done to Earth could ever be patched up. Petty regional interests, our tribal instincts acted out and legitimised selfishness seem to win out every time. It won't always be this way, once our food sources are rationed, once the water literally dries up, when electricity and broadband switch off - the riots will hit the streets. If I'm right of course, it's too late by then anyway. It doesn't mean we shouldn't try, those of us who care anyway, but it does have me sighing wearily and reaching for another Pina Colada, at least while the humble pineapple still exists anyway.

Try as we might, which is to say we don't try that hard at all, we cannot seem to help ourselves from filling up the ocean with crap that hurts it, and by extension us. Take microplastics for example, which you can probably ascertain from their name are small pieces of plastic. They float free in the ocean, and are often ingested by marine organisms, causing them no end of harm. They gravitate up the food chain, indigestible, until they arrive on our own plates.

Where do they come from exactly? Nature, in its illustrious several billion-years of history has never evolved plastics. There are some natural polymers out there including wool and silk, but those aren't choking the oceans, so we can't point the finger in their general direction.

One of the primary sources of microplastics is the Great Pacific Garbage Patch, an area of increased plastics density between Hawaii and California. Swept together by ocean currents, this vortex contains an estimated two trillion individual pieces of plastic discarded by humanity. There's no precise data available for the origin story naturally, but sources like industrial waste and discarded fishing equipment are believed to be the primary sources. It sure as hell isn't the dolphins poisoning us, though that would make for a magnificent conspiracy theory.

Or there are the jellyfish. A humble squidgy creature of the deep, their population has exploded in recent years, brought on by overfishing of the creatures that would otherwise feed on them, and climate change making the oceans warmer and more acidic, just the way they like it. While themselves are relatively harmless (though you wouldn't say that if you'd been stung by one) jellyfish pose a hazard bizarrely to nuclear power stations, which always need to be built near a coastline because of the cold-water intake needed to regulate nuclear fission. It doesn't take more than a few hundred jellyfish to clog an intake pipe. Meltdown, anyone? They also present a significant hazard to fish farms, where they can consume the oxygen needed for the farmed salmon, which can hardly swim away to a more oxygen-rich area while caged up.

While tackling the root cause would make a certain amount of sense, the EU, with a certain sense of resignation has tasked scientists around the world with finding ways of using these excess jellyfish in some way. The crazier suggestions include cosmetics, since they are a naturally occurring source of collagen. Next time you use a face cream, imagine how much softer and nicer it would feel, knowing it came from our squiggly cousins of the ocean, not to mention injecting them into your face to tackle wrinkles and blemishes.

Even beyond the aquatic life forms that we just love to consume, the oceans provide us with another, even more basic and primary need, the water itself.

In 2018, a study by the University of Queensland found that 87 percent of the world's oceans have been contaminated in some way by human impact, leaving just some remote areas of the South Pacific as well as the Poles relatively unspoiled. The oceans cover 70 percent of our planet, which makes our insanely destructive efforts oddly impressive in their own perverted way.

You may recall natures elegant water cycle. Most of us were taught it at school, which by way of a refresher, is where sunlight heats the sea, turning a tiny fraction to water vapour which then condenses into clouds to fall as rain over land, which then runs down into the sea. I've oversimplified of course, but I'm very good at that. In any case, somewhere in that cycle we have jammed a tap in and are guzzling all the water we can get our mouths on.

And yet, despite all this knowledge, freely and publicly available, as a species we just can't stop ourselves from poisoning our own watering hole. If humanity had one greatest failing, above all the other whopping great ones, it would be our failure to act in our own collective best interests. The reason behind this failure of course is because we are all still tribal at heart, and with that, still monumentally selfish. In our colossal arrogance, we believe ourselves far removed from our primitive beginnings, yet we work individually, act locally, think regionally and behave provincially.

There are exceptions of course, many scientists have risen above these simple beginnings, yet this leads them to operate as a clan in their own right, a clan often ostracised by the alpha males (aka congenital imbeciles) who often seem to wind up leading our regional tribes. Oh,

Global Warming is a problem? Blame the scientists. Maybe they'll learn their lesson after we cut their funding. I should point out once more, it's not that males have exclusivity rights on stupid, short sighted behaviour, but in this patriarchal society it's more often than not the case. More on our patriarchal society later on.

The language, political and economic barriers we've put up all around ourselves are all the evidence we need that we haven't escaped our tribal beginnings. That, and the fact that repairing a broken, dying world is always, always going to be somebody else's problem.

# Part 2 – How We Met our Tipping Point

We are very, very small, but we are profoundly capable of
very, very big things.
Stephen Hawking

# CHAPTER 5 - THE HOLOCENE EXTINCTION

The Holocene Extinction. Sounds like a good disaster movie
doesn't it. Also known as the Anthropocene Extinction, also the Sixth
Extinction. I'd go and see any of those at the cinema on a slow Sunday
afternoon too. As you may have gathered from the context, this relates
to an ecological die-off, a period where great swathes of organisms
quietly shuffle off the mortal coil. What differentiates it from the five
that came before it, is that it's still going on. Right. Now.

Where to begin? Well, to start with, the Holocene is the geological
period we currently inhabit, taking the baton from the Pleistocene,
when things were a good deal colder and living conditions were way
less comfortable. Most of Earth's biologists (more than 70 percent of
them according to a survey from around 20 years ago) are convinced
that the Holocene Extinction is a real thing, it's happening right now,
and worst of all, it's only just getting warmed up.

This extinction of which I speak started at very beginning of the
Holocene, the transition into which in turn was more or less at the end
of the last major ice age, which in our calendar was about 12 thousand
trips around the Sun ago for our fair planet.

It all started with the extinction due to the intense cold of various megafauna, which for the semi-uneducated (like myself) means, bloody big creatures - the Woolly Mammoth is one you've doubtless heard of, but others include larger breeds of sloths, buffalo, rhinoceros and crocodile. While on the subject of megafauna, one which caught my eye found its extinction a good deal earlier in the Permian-Triassic Extinction Event 252 million years ago was a creature called the Meganisoptera. Not so catchy is it.

How about METER LONG DRAGONFLY??? I bet that got your attention, it certainly did mine. Just imagine the size of the rolled-up newspaper you'd need for that buzzing horror show. And don't get me started on the Arthropleura, a nine-foot long millipede with pincers as long as your forearm. Nope. Just nope.

The commencement of the Holocene period around 12,000 years back and the related extinctions spookily enough correspond exactly with the migration patterns of a certain tribal species with an unhealthy fascination for annoying reality TV and even more grating social media. Yep, this was the start of humanities hostile takeover of the planetary ecosystem, when we really got our act together, such as it is. After so many false starts and blind alleys, we started to spread out and make ourselves comfortable, with no regard for, or even awareness of an ecosystem around us that might strangely enough have some limits.

An so it was that humanity became Earth's first 'Super Predator', the first hunter species that was fully capable of hunting every adult alpha predator on the planet and mounting it on their wall. A lethal killing machine capable of endless ingenuity and savagery, that billions of years of evolution had never once produced before. Not the Tyrannosaurus Rex on land, nor the Megalodon giant prehistoric shark at sea could hold a candle to the determined raw extermination potential that humanity wielded, and all creatures fell before their callous hands.

They do of course continue their chaotic abuse of the ecosystem to this day, with nothing capable of standing in their way, least of all their own self-control or better judgement. The species that have fallen beneath the spear/knife/blunderbuss of these super predators is incalculable, in that it's far too late now to try and calculate them.

Darwin and his naturalists would have had to start work several millennia early and buy some really big filing cabinets to keep track of all the poor flora and fauna that would depart this earth forever at the hands of the super predators. They barely know they are doing it half the time, so uncatalogued are the more remote species that still feel the impact of changes to the environment, the plants, the animals, the water, to the very air itself.

Here then, we arrive at the sad case of the White Rhino. To be more accurate, I'm talking about the Northern subspecies just in case you were feeling pedantic. The Southern White Rhino is amazingly almost doing okay, barely even going extinct, especially by comparison to their poor Nordic cousins.

Speaking of being pedantic, an old friend of mine long since wired and shackled to a future very different to mine once presented me with a classic conundrum when he told me to not be so 'pandantic' when I was doubtless being obnoxiously specific about something, as I apparently have a tendency to do. The urge to correct him was almost irresistible, and exactly, equally opposite to the desire to not prove him right. To this day I couldn't tell you which was the right decision, but I imagine committing the debate into a book may count as having found a third option.

Back to the rhinos, let us begin with a quick headcount. At the time of writing this section (early 2019) there are just two Northern White Rhinos alive in the world. Not just in the wild, not just in captivity; anywhere on the entire planet, except in the past if you were feeling temporally facetious (not to be confused with temporarily, or faeces for that matter). The two surviving rhinos are both female, so there is little chance of a consummation that would lead to a child, and even if they were male and female, and even if they took a shine to each other, we'd still then only be talking about three left. That would be is great if we were talking in statistical terms (a 50 percent growth in the population to save you doing the math), but still not so good in terms of the genetic variety needed to safeguard a species, which is more in the many dozens or preferably hundreds.

All of which means that for the Northern White Rhino, unfortunately and tragically, that ship has well and truly come, gone, sailed away and then sunk.

So, what actually-on-Earth happened? I've always found that a superfluous statement 'what on Earth' amusing, even though I do use it myself to avoid particularly religious-sounding epithets, when I'm not feeling subversive enough to drop the flying spaghetti monster into conversation. My take on it is, pretty much everything can be considered 'on Earth' unless you happen to be on an orbital space station or technically very deep underground when you'd technically be 'under Earth', and I bet it's been uttered in those localities also. Oh well. I also have to allow that it may well, linguistically speaking be a contraction of "What on God's Green Earth", which I'd have to concede makes my obsessive epithet avoidance amusingly superfluous.

The Northern White Rhinos natural (and literal) stomping ground for tens of thousands of years ranged across thousands-of-kilometre long swathes of central Africa, across the borders of what are now Uganda, Chad, Sudan, the Central African Republic and the Democratic Republic of Congo. The human residents for much of that time stayed out of their way if they had any sense, which was prudent given they have a fearsome reputation for being bad tempered, coupled with a sharp horn and an average body weight upwards of a metric tonne.

In our very recent human history though it all started to go wrong for the Northern White Rhino, two things happened more or less simultaneously (at least from a geological perspective) in the 1970s and 1980s, that had simply disastrous implications for natures noble, naturally armour-plated killing machines.

Firstly, if there's one thing that ironically unifies the recent history of those previously listed countries it would be their respective geopolitical instabilities, which have led in many cases to little or no governments during the time in question. Secondly, as China and much of Asia slowly became more and more developed, dragging themselves painfully into what passes for the modern age, more and more people could afford more and more things. While this counts as standard consumerism for the most part, the demand for rhino horn, long used in Chinese alternative medicines especially concerning aphrodisiacs and virility went through the roof. I find this a special and terrible irony that such ancient (and most certainly wildly incorrect) homeopathy would find increased purchase in the modern age, but this is the tip of the iceberg when it comes to the idiocies perpetuated by humanity.

The clever and sublimely funny comedian Tim Minchin once castigated homeopathy in the most perfect of ways, simply by pointing out that all so-called alternative medical treatments eventually are either proved to work or proved not to work. If they are proved to work, all we do is drop the 'alternative', and just call them medicine. If not, they remain the province of nebulous homeopathy and fake reviews on the internet.

Our two causative factors then were perpetual conflict and rising demands for Eastern medicine. While either one of these factors by itself wouldn't have dramatically affected the rhino population, the two combined led to an explosion of intensive, unmonitored killing with no regard whatsoever for the number of animals that might left alive. I couldn't even use the word poaching to describe the situation, as when this all started in the 1970s, it was by no means illegal. At one point the governments of that part of the world even monetised it by selling licenses to commit this unspeakable act. Even when it did finally become technically illegal, little was done to deter the poachers, even today the anti-poaching teams are underequipped, underfunded, and often shot at by the morons who see no issue with depriving Africa of some of its oldest and most interesting species.

And so it began. The Northern White Rhino, which at one point enjoyed a stable population in the high tens of thousands entered a critical decline from which it has never recovered.

The last male of the species, named Sudan died in March 2018, leaving behind the final two females, Najin and Fatu to fend for themselves in the Ol Pejeta Conservancy in Central Kenya. They are under round the clock protection, as even just the two of them sadly eyeing each other wondering what the hell happened is an attractive target for poachers. In fact, back in March 2017 a group of 100 percent certified awful human beings sneaked into a zoo in Eastern France and killed a Southern White Rhino for his horn. If the universe possesses any karmic justice at all, those despicable excuses for people will all trip and fall onto some very sharp and pointy things at the very earliest opportunity. The universe doesn't have any such self-correcting mechanism of course, but one can still entertain such daydreams and in the face of such stupid and pointless cruelty, it at least beats going mad.

Genetic engineering has long been touted as a possible saviour to species like the Northern White Rhino facing extinction, perhaps infusing the ova of another, less critically endangered species of rhino with Northern White Rhino DNA, thus resurrecting the species from the very brink of extinction. Under the banner of 'Facilitated Adaptation', scientists hotly debate the pros and cons of assisting the endangered species of the world to avert their demise. While noble, this area of scientific research is very much in its infancy with many pitfalls doubtless awaiting us, and little is known about the long-term risks involved in such an effort.

One way Facilitated Adaptation would benefit a species is to allow them to better weather the changes in their environment, e.g. jungle to desert, rainforest to swampland. While the environment has certainly changed on the Northern White Rhino, it's not the primary cause of their extinction; we hold that privilege, to our eternal discredit.

In any case, let's say we succeed, that the genome of the Southern White Rhino is successfully sequenced into the Northern White Rhino genome in order to produce an 'adapted' Northern White Rhino that inherits the considerably better survival traits of its southern cousin. While I concede that such an endeavour could take place and even succeed, it's unlikely to do so in the lifetime of Najin and Fatu so, for the immediate future at least, the Southern White Rhino is destined to become extinct. Another sacrifice at the altar of 'sucks to be alive and not human on this planet'.

The Southern White Rhino stands as a poignant symbol of all extinctions caused by man, now and yet to come. In the same survey I mentioned earlier, the majority of biologists agreed that 20 percent of all living species could become extinct by 2028, a figure we appear to be well on course to exceed. The rampant trade in ivory continues to this day. Mozambique in particular suffers from one of the highest rates of ivory poaching: since 2014 more than half its elephant population is believed to have been killed by poachers.

Hunting of course is merely one of the mechanisms human beings use to depopulate the planet of so many of our fellow residents. I touched on climate change a while back, where our abuse of the ecosystem tips the climate into a new configuration that brings about reductions in the population of other species, and in the worst case, wholesale extinctions. The punchline of the cruel joke we play on the planet will be the final species whose extinction we bring about.

More on that later.

Also, in case you thought I'd somehow missed it, destruction of habitats for our agricultural requirements is another biggie – so big in fact that it's got its own chapter coming up for your delectation.

Disease is another causative factor of the Holocene Extinction, many of them wrought by humanity, mostly by accident but very occasionally on purpose. Human beings carry diseases everywhere they go, something that you quickly realise when commuting in London and spend nine months of the year nursing a cold. Our aboriginal ancestors walking out of Africa brought new diseases with them, diseases that frequently hop, skipped and jumped back and forth between species, decimating entire regions. As these diseases evolved and adapted, the travelling humans would ferry them back again, reintroducing the adapted version to a region that had no defence, starting the circle again. 'Travel is fatal to prejudice, bigotry, and narrow-mindedness' as Mark Twain once famously said, and of course he was absolutely correct, but it sure does help with the spread of infectious diseases.

Another more recent example concerns the Variant Creutzfeldt–Jakob disease (vCJD), which hit the headlines in the UK with splash in the late 90s. vCJD is a terrible brain disease that occurs in humans, and is caused by eating tainted beef, specifically beef that is infected with

Bovine Spongiform Encephalopathy (BSE), or its more colourful moniker, Mad Cow Disease. In cows, BSE causes a degradation of brain function, including ability to move, weight loss, abnormal behaviour. Everything you might expect from a mad cow funnily enough.

The ultimate cause of BSE going from a random biological occurrence to an outbreak turned out to be the extremely dubious practice of feeding cows on the remains of other cows.

While cannibalism turns out to be surprisingly commonplace in the animal kingdom, that generally involves eating another creature who has very recently died. The flesh was fresh; if like me you enjoy an alliterative sentence. In the case of the poor cows of Great Britain, they were being fed meat-and-bone-meal (MBM), which is a product of the rendering industry (take my advice - do not google that), and is a charming cocktail of proteins, ash, fat and water of their fallen brethren, and certainly not the kind of thing that shows up spread on crackers at your average food festival.

MBM has since been banned globally for feeding multi-stomach creatures, where the fermentation process they undergo as part of digestion cooked the MBM up nicely into a fresh batch of BSE, then when those cows were eaten by humans, brought about the dreaded vCJD.

At other times in our history our assistance in spreading disease has been even more deliberate. As we covered in previous chapters, our species evolved in Southern Africa a few million years ago and expanded from there across the land bridge into Europe, and from there spread around the planet. Wave after wave of migration occurred, blind alleys were frequent as the wax and wane of ice ages and climate changes wiped out entire branches and generations of our ill-prepared ancestors, who could scarcely have understood what a wall of ice suddenly encroaching on their territory might mean.

After the end of the last ice age, the disparate descendants of those African hominids banded together into larger and larger units. Civilisation, with all its attendant problems started to spring up in various places. It was hardly a uniform experience, while some groups of humans began to develop technologically and culturally, other groups did not, instead retaining rigidly static cultures down through the ages.

The groups of humans who developed further started to live longer and to grow their populations. For that they needed space and lots of it. Europe around the 14th Century is a prime example of this, but it was by no means the first. Explorers sailed to the four corners of the world, no easy task when the planet is spherical, but still they managed it. They found and brought back wonders, but no wonder was as great as the prospect of all that new living space, only encumbered by some native populations who were hardly a threat to the weapons and tactics that being cooped up in Europe for all those centuries had brought about.

The groups of humans who over centuries had developed into complicated yet largely meaningless partitions known as English, Spanish, Portuguese and French (among many others) took quick advantage of this new land area to expand their people and cultures into. North America alone has almost ten times as much land area as England, Spain and France combined, and was at that time sparsely populated. It was to become even more sparsely populated, as the advancing colonials brought with them smallpox, measles and cholera, to which the native inhabitants had no defences whatsoever. When the colonial forces couldn't subdue with their primitive chemical warfare (which included deliberately trading smallpox infected blankets with the Native Americans), they depopulated by more direct means, like the Wounded Knee Massacre (sometimes incorrectly called The Battle of Wounded Knee, but make no mistake, there was no battle here). At this horrific event more than 250 unarmed men, women and children were killed by soldiers from the United States 7th Cavalry Regiment. More than 20 of the soldiers were awarded the medal of honour for their disgusting actions. As a species, it was hardly our finest moment, and yet still merely a footnote in the annals of a history replete with murderous actions where the greedy and strong preyed on the innocent and weak to take whatever they wanted.

There is a prominent school of thought that capitalism is ultimately to blame for all this destruction, and indeed if plotted on a graph, there is a disturbing correlation between capitalism becoming the dominant global economic force (replacing mercantilism) and the acceleration of species extinctions. I suspect the root cause is deeper than that, however. China, late to the industrial revolution party is fast becoming the prime emitter of greenhouse gases, a major contributor to climate change and species extinction, and capitalism can hardly be applied to that socialist 1984-style thought policed megapolis.

Earlier on I referred to the Holocene Extinction as the Sixth extinction. Then what happened to the other five, is an obvious question that bears shining a little light on given our current predicament. First off, it's worth exploring how we identify that there was an extinction in the first place, these go back some time in history so there wasn't a paleontological equivalent of a train-spotter leering on the end of the platform for the 8.38 from Doncaster. Or was there?

The mechanism for identifying an extinction after the fact is by looking at the fossil record, the complete, organised, time-indexed log of every fossil ever found. While it's a complete log of all fossils, it is by no means complete – huge gaps abound where fossils have yet to be discovered, if indeed they ever will, given that for the most part only things with hard bones or shells survive the uncounted millennia till we dig them up. Despite the picture of our history being imperfect and incomplete, the dated points in the record where creatures go from being there to not there tell us something important too – a lot of stuff died. Given we're only looking at the smallest sliver of what creatures were actually alive, one can only imagine the millions of species who for whatever reason didn't make it to the fossil record yet died off just the same.

In order of appearance, the big five extinctions that have come before are as follows.

First, we have the Ordovician–Silurian extinction, which occurred 444 million years ago, and resulted in 86 percent of living species at the time going extinct. The cause is thought to be an ice age, potentially triggered by the uplift of the Appalachian Mountains, which was going on at the time, the vast volume of newly exposed rock would have removed a lot of $CO_2$ from the atmosphere, reducing the temperature of the planet and allowing the glaciers to take hold.

Then came the Late Devonian extinction around 375 million years ago, which was slightly less brutal, and only killed off 75 percent of living flora and fauna. This was around the time of the first plants, so it's theorised that they may have had something to do with it, as Earth adjusted itself to its new inhabitants, the atmosphere itself changing dramatically for the better (speaking for the species to come in any case). This in turn then may have triggered algal blooms, starving the oceans inhabitants of much needed oxygen. The most advanced creature in the ocean at the time was the Trilobite, whom evolution had gifted with complex eyes and a tough hide for fending off predators.

The next one was as they say, a doozy. The Permian–Triassic extinction event 251 million years ago goes by the more sinister title, 'the great dying', and as nicknames go, it's very much on the money. 96 percent of the species alive at the time on Earth died off, which is the closest life on this mudball came to being completely over before things even got interesting. The cause is believed to be a super-volcano eruption in Siberia which sent enough material into the atmosphere to trigger a nuclear winter.

All the trees in the world died. All the coral in the world died. The trees and coral that exist in the world now bear no direct relation to them, they had to evolve again from earlier, more hardy forms that were better able to weather the storm. Palaeontologists estimate that this volcanic eruption set back the evolution of life as we know it by more than 300 million years. Imagine what we would have by now if this hadn't taken place.

Stick with me, we're on the home stretch now. By the way, in case you are wondering why the extinction events are double barrelled, it's because these major extinctions actually mark the boundaries between these defined geological ages, which gives you an indication how significant they were.

Next up is the Triassic–Jurassic extinction event 200 million years ago which saw a loss of around 80 percent of the species known to exist at the time. I hope you like mysteries because nobody knows yet what caused this one, no asteroid impacts, volcanic eruptions or major climate shifts correlate with the time of the extinction. Hopefully like me you're used to life's little disappointments and you can move past it without it keeping you up at night.

The last historical extinction, not including the one we're currently both experiencing and causing is the Cretaceous–Paleogene one, which happened just 65 or so million years ago. This extinction put an end to the reptiles better known as the dinosaurs. The dinosaurs had been stomping around since the Triassic period, around 240 million years, and managed to survive the last great extinction, but their luck was about to run out. An asteroid tens-of-kilometres across impacted the Yucatan Peninsula. Another nuclear winter followed where the majority of the plants, with no access to the sun's rays died off. No animal greater than 25 kilograms survived. Not all the dinosaurs died though, interesting enough. The smallest and most agile among them lived through the winter, probably due in large part to the feathers that covered them. Their descendants are alive today, split among around ten thousand species, number approximately 400 billion and wake me up annoyingly early, especially after I've been on the red wine. I am of course talking about the birds.

Ebb and flow, the species of Earth, on these cosmic timescales rise and fall in the blink of an eye. These are also just the extinction events we know about, the further back you go, the sketchier and less accurate the fossil record is, it's entirely possible, probable even that a great many other extinction events were experienced, survived by just a handful of the species. And that's the important thing, each time an extinction event happens, the hardiest survivors go on, proliferate, evolve, diverge and converge. New stronger species from the ashes of the old.

And so it will go for the sixth extinction. Some species will survive the changing climate, and their descendants will evolve into wholly new and interesting forms. I doubt humans will be one of them. A recent study put that one million (out of eight million) species are facing extinction as a result of our activities, I'd be very surprised if the true number wasn't much higher. Another study concluded that at present rates of decline, insects will vanish from Earth's surface within a century. Many of these species depend on one another for survival, the complex ecosystem with its delicate balancing act can only take so much abuse before it all comes crashing down, a collapse so slow nobody will notice until it's far too late.

Down the line, some of the surviving species may be intelligent enough to become truly self-aware, maybe even develop language, a culture of their own. I wonder what they will make of the fossils we leave behind.

# CHAPTER 6 – ATMOSPHERE

The atmosphere. In our subjective terms, the very air we breathe. It is without a doubt the most fundamental of human rights, one that we take for granted more than any other, until there isn't any more of breathable quality left at least, where it may creep up the priority scale somewhat. The atmosphere is closely linked to the oceans through the water cycle; as previously established, they can be thought of as two equal opposites, yins and yangs, with the land we occupy sandwiched awkwardly in between like an unappetising filler. The marmite, if you will.

Going way back, the original atmosphere of Earth was very different of course. In the earliest of early days, around 4.5 billion years back it originally consisted of hydrogen, water vapour, methane and a few of the simpler carbon oxides. The source of this original atmosphere would have been Earth itself, as the interior of Earth cooled and settled, it underwent a phase known as degassing, not unlike the period at your parents' house after a hefty roast dinner. These lighter gaseous elements that comprised part of Earth rose upwards, not far enough to get into space, they still had gravity to contend with, but they left the rock strata and migrated northwards, eventually pooling all the way around the outside of this unremarkable ball of until recently molten rock. One key element of the atmosphere that we know and love today was missing of course, oxygen.

As touched on when we explored the origins of life, oxygen turned up to the party around a billion years ago when the little oceangoing critters, our earliest ancestors, developed the ability to metabolise light: the process of photosynthesis. In one of nature's most clever evolutionary innovations, these creatures organically developed the mechanism to use energy from the sun to split molecules of water ($H_2O$) and carbon dioxide ($CO_2$) and recombine them into organic compounds they could use. The waste product of this fascinating and literally game changing development was oxygen ($O_2$). Being substantially lighter than water, the oxygen ascended, accumulating in the envelope of gas that surrounded the planet.

What happened next could not have been foreseen, and nonetheless is another of the happy (if you want to call it that) accidents that allowed life to flourish in the way as it has on our mudball. High up in the atmosphere on the edge of space, the oxygen soared free, at this altitude the ultraviolet rays of Earth's sun, a lethal sleet of exotic particles generated by ongoing fusion reactions interacted with the oxygen. This high energy interaction caused the split apart of some of the oxygen into single oxygen atoms, and their subsequent binding with the surviving $O_2$ created something wholly new and extremely helpful. Quick maths test for you, what is $O_2$ plus O? I am referring of course to $O_3$, which also goes by the street name Ozone.

The penny starts to drop no doubt. At the upper limits of our atmosphere, where something suddenly becomes nothing in a very big way the ozone, continually replenished by the sun's death rays absorbs a significant proportion of the harmful ultraviolet radiation which had for billions of years rained down upon the uncaring surface of this world. Without this outer layer of ozone, it's questionable whether life would ever have emerged from the seas, which are naturally protected from the rays, since interestingly enough water is an excellent radiation shield. Just a few metres of water were enough to provide the difference between life and death for our early ancestors, a trick which was not lost on designers of nuclear power stations, which make good use of water to avoid disastrous contamination.

Suffice to say, organic matter and ultraviolet radiation do not good bedfellows make. This invisible shield around Earth has been in place now for a billion years and is the sole reason you don't get sunburned every single time you step outside for more than oh, three seconds, though for a few of my paler friends that is still the case.

It almost makes an interesting debate point on the topic of intelligent design, in that so many random events and unlikely happenstances took place in the billions of years that drifted by before humanity even started speculating about them. If any of these quirks of fate had happened differently, the most advanced denizens of the planet today would likely still be confined to the sea. Perhaps in their advancement they would happen to swim faster than others, react faster than others, but their rudimentary thought would still be restricted to the basics; food and mating, fighting and flighting. Maybe not as much has changed as we thought. Certainly, there would be no room for abstraction, or even consciousness by any recognisable metric. I did say almost, as random chance is still infinitely more likely than a divine creator, though I have always admired the somewhat inaccurate analogy deployed by intelligent design about random evolution: it's as if a tornado swept through an aircraft hangar full of parts and assembled a 747. I find it an elegant analogy if highly questionable in its reasoning.

Around forty years ago in the late 1970s, scientists however discovered something shocking about Earth's atmosphere. Tests appeared to show that total amount of ozone in the atmosphere was decreasing, and by the mid-eighties it was clear that there was actually a hole in this vital coverage over the Antarctic and it was growing bigger by the year. At this point you may be thinking 'bloody CFCs' and shrugging supportively. And it's true, we all know the story, but the truth is rather darker than you may think.

As far back as 1973, scientists were aware that chlorofluorocarbons, a hydrocarbon used extensively in the chemical and refrigeration industries could be causing some harm to the ecosystem. If you know anything about human beings, you know the scientists were immediately taken seriously and actions quickly taken to rectify the damage we had inadvertently caused. Well no, of course not, quite the opposite as is our way.

It all started with James Lovelock, wacky scientist and environmentalist. You may have heard his name connected with the Gaia Hypothesis, the proposal that Earth and all the living creatures that inhabit its surface are all part of a single, albeit inordinately complicated system, a system which self-regulates in a semi-conscious manner to perpetuate life on this world. Oddly enough he's incidentally a big figure in cryonics also, having successfully frozen and thawed out lab rats. And by successfully thawing, I don't just mean they stopped being frozen, the rat was alive afterwards, which in cryonics is something of a high bar. I doubt it had the mental faculty to compose a symphony orchestra after (if indeed it could before), but it opened the door to some intriguing possibilities around life preservation beyond the normal human lifespan which aren't relevant to this discussion, but I may well write about them some other time.

Back in 1957, Lovelock developed the Electron Capture Detector, a clever little device designed to detect which atoms or molecules occur in a gas. It does this through a process called Electron Capture Ionization. This wasn't the simplest thing to get my head around, my grasp of chemistry beyond the construction of a gin and tonic is at best rudimentary, but I shall attempt to explain. The ECD uses a radiation source (typically a beta particle emitter) which shoots electrons through the gas in question. Where said electrons collide with atoms and molecules within the gas, many additional electrons are created. These free electrons are then drawn to the other end of the ECD chamber by an electrically charged anode, and the concentration of electrons allows deductions to be made on the composition of the gas.

At the time, this took the world of gas chromatography, already one of the trendiest branches of chemistry, by storm, proving even in 1957 to be approximately 10,000 times more accurate than its nearest rival. In the late 60s using this technique, Lovelock discovered the concentration of CFCs in the air over Ireland to be around 60 parts per trillion, but since the prevailing understanding of the time was that the air should contain zero parts per trillion, it was something of a cause for concern. Then, in 1972 Lovelock took part in an exhibition he himself part funded, on board the RRS (Royal Research Ship) Shackleton which took him from jolly old England all the way down to Antarctica. He took readings with his Electron Capture Detector every day during the trip, and for the first time a global picture concerning the persistent presence of CFCs on our atmosphere (which appeared in every single sample) began to form.

However, something untoward happened here, something I'm sure still keeps old James, at the time of writing currently at the magnificent age of 99, up at night. He concluded that CFCs constituted 'no conceivable hazard', much later correcting himself to mean 'no conceivable toxic hazard', but the damage at the time was done, with the incorrect conclusion that there was no damage being done. As I mentioned though, the discovery of CFCs in our atmosphere started with Lovelace, it did not finish with him. Thanks to his work though, it was at least commonly understood that manmade CFCs were present in our atmosphere, a fact which could be built upon by the next generation of scientists.

Enter F. Sherwood "Sherry" Rowland and Mario Molina, both experts in chemistry. After attending a lecture on the subject of Lovelock's discovery about CFCs permeating our atmosphere, it got them thinking about whether there could have been any unforeseen implications of said chemicals freely floating around our little shell of air. By 1976 they had managed to prove that not only was the amount of ozone in the atmosphere depleting at a rate never before seen, but also managed to connect CFCs to it as the direct culprit. It turns out that one of the most useful qualities about CFCs, their low reactivity with other chemicals, turns out to be one of the most dangerous once it hits the atmosphere. Since this lack of reaction give them a lifespan of more than 100 years, they have all the time they need to diffuse into the upper stratosphere, which is where things start to go wrong.

Chlorofluorocarbons are a chemical compound consisting or, you've guessed it, chlorine, fluorine and carbon, and this chemical combination also likes to go by the stage name in North America of Freon. CFCs are emitted by our civilisation as a waste product from discarded fridges, landfill and the like. Once they are into the upper reaches of our stratosphere, the intense ultraviolet sunlight causes the CFCs to break down into simpler forms and release Halogen atoms through a process known as 'photodissociation'. Halogen in this context is extremely bad, as it catalyses (speeds up a chemical process) the natural breakdown of ozone back down into regular oxygen, tipping the delicate balance to the point where more ozone was being broken down into oxygen courtesy of the CFCs than was being created by the ultraviolet bombardment. By the time it was noticed, the volume of ozone in the layer had reduced by more than 4 percent, which given it had taken barely 50 years for humans to cause, has to be some kind of record for self-inflicted habitat damage.

Of course, CFCs didn't go down without a fight. There was way too much money on the table, at that time in the 70s the chemical industry was worth tens of billions, and no pesky inorganic molecule lurking at the outskirts of our atmosphere was going to get in the way of that. In a scenario that played out much like the tobacco lobbying of previous decades, the chemical industry founded such laughable official sounding institutions as the Aerosol Education Bureau, the Council on Atmospheric Sciences and the Western Aerosol Information Bureau. These agencies employed a range of despicable dirty tricks ranging from outright denial, accusations that regulation of the industry would amount to Communism and good old-fashioned smear campaigns against the noble scientists trying to stop humanity from giving itself a terminal case of skin cancer.

Their actions succeeded in delaying the implementation of a ban on CFCs for more than a decade before the Montreal Protocol on Substances that Deplete the Ozone Layer came into force on 16th September 1989. I hope they are proud of themselves, and depressingly, I rather expect that they are.

This is part of the endemic problem of our current human society. These things, everything I write about in fact are just not our problem because the issues aren't affecting us on any scale we can perceive. They might affect our children, and our children's children, which surely given our Wirings maternal and paternal leanings to force us to care for our children should prevent the worst from happening, but somehow, it just doesn't. That fact, coupled with our cultural bias towards those with the least empathy gaining the most power, money and influence unfortunately prevent us from progressing any further forward as a species.

But I digress. To be fair, the Montreal Protocol has been ratified by all 197 member states of the United Nations, which in 1989 was something of a first, and depressingly the fact that it only took 14 years to agree was also hailed as an incredible achievement of international cooperation. Since then the only United Nations treaties also ratified by every country involved follow a certain theme, the UN Framework Convention on Climate Change and the UN Convention to Combat Desertification, both closely related to the concerns I'm trying to convey.

With CFCs banned, the Ozone Layer is slowly recovering. Scientists estimate that it will be restored to its pristine level in another 30-40 years and until then it is advisable to wear a hat and carry a parasol should you find yourself out and about in certain parts of the Antarctic. In recent years however, a disturbing trend has seen a substantial decline in the speed of reduction of the CFCs in the atmosphere since 2012. It isn't so bad that the reduction trend is reversing just yet, but it's clear that someone is pumping out CFCs once more.

Would you care to hazard a guess? Who would be so mind bogglingly stupid as to dramatically ramp up our species' chances of getting skin cancer, all in the name of saving a few bucks? None other than our old friend China. The Communist Party, wrist having been slapped by the Environmental Inspection Agency has promised a clamp down on said production. China were a signatory to the Montreal Protocols of course, they've even gone so far as to arrest a few 'rogue individuals' in 2018, but as for whether these individuals were acting on a state sponsored basis, we shall likely never know, nor will the individuals ever be seen again.

The scary volume of CFCs in the atmosphere is also just the tip of the air-berg, we've yet to touch on greenhouse gases, the biggest bear of them all, atmospherically speaking. Greenhouse gases are gases which absorb energy, the main ones in our own atmosphere being water vapor, carbon dioxide, methane, nitrous oxide and ozone. I use 'our own atmosphere' specifically, as Mars, Venus and even Titan orbiting distant Saturn also contain greenhouse gases in abundance.

The word greenhouse is, as always in popular science used

inaccurately here. An actual greenhouse, where your beloved nan might have forced endless egg and onion rolls down your neck as a child (too specific?) operates by preventing the flow of air heated through the glass panes by the Sun, thereby retaining a stifling heat. Greenhouse gases function by permitting the light energy from the Sun through them, but then not permitting the heat energy created by said light hitting Earth's surface from escaping again. It may be a subtle difference, but I try to be the best kind of correct wherever I can.

Greenhouse gases are bad right? The news is always going on about them in doom and gloom articles after all. Shouldn't we be doing all we can to stamp them out?

Not so.

Greenhouse gases are in fact extraordinarily useful, in that they trap the Sun's rays within their protective envelope, and that energy absorption consequently warms the surface of Earth. Without greenhouse gases it's doubtful that life would ever have emerged from the seas. The development and subsequent explosion of single then multi-celled life in the seas would have happened just the same, but without the warming effect of the greenhouse gases multiplying effect of the Sun's radiant energy, it's estimated that the median temperature of our planet's surface would be a brisk −18°C (0 °F).

We've established then that we need greenhouse gases more than we thought we did, but as with everything in life (even red wine), you can have too much of a good thing. It is estimated that human activities on the planet in the last 250 years, since the Industrial Revolution started us on the slow, inexorable path towards doom, have increased the amount of carbon dioxide from its natural 280 parts per million to 407 – an increase of 45 percent. Fuelling this increase somewhat literally, is the burning of fossil fuels, which we shall come onto in due course.

You can't have missed the messaging, warning after warning put out by the world's scientific community, desperate pleas to cut our carbon emissions before we increase temperature irreversibly. How our actions will cause the melting of the polar ice caps, in turn causing sea levels to rise and drowning forever those coastal areas that seemed so attractive to our earliest ancestors as they stepped off the plains of Africa. How the rising temperatures will wreak havoc amongst crop yields on our remaining arable land, the land which isn't now under metres of water in any case.

Those pleas of course are going unanswered, since as ever, our lack of global cohesion and the ability for certain soulless individuals (primarily old white men with a lifetime of callous behaviour behind them) to make these momentous decisions on our behalf. More on this very soon.

Carbon dioxide, while being the major contributor to greenhouse gases is hardly the only culprit. Methane, for example is another bad one. While it makes up just 14 percent of greenhouse gases, it is 28 times more efficient than carbon dioxide at trapping heat in our atmosphere. A huge contributor to our excess methane is livestock, the animals we have been rearing for hundreds of years to sustain us, primarily cows, but also including pigs and chickens; so we can drink their milk (only cows mind you – I don't think pigs milk will be gracing the shelves of supermarkets any time soon) and of course, consume their flesh.

I'm acutely aware incidentally also that I am referring to gases as the main culprits, but on the whole, I don't think we can say it is the carbon dioxide's fault that there is a disproportionate amount of it in our atmosphere, can we.

According to the Food & Agriculture Organisation (FAO) there are just under 1.5 billion cows in the world. To be fair the websites concerned state there are 1.5 billion head of cattle, but since it (a) amounts to the same thing, and (b) sounds a lot less silly, that's what I'm going to go with. That's a lot of cows, and more importantly, a lot of cow's arses. Yes, that's right, a primary contributor of methane into the atmosphere where it works its dangerous magic as a greenhouse gas is cows farting.

As you may be aware, cows have a rather spectacularly complicated digestive system, consisting amazingly of four stomachs, which ferment the plant matter into a digestible form. Human beings, with their somewhat simpler digestive systems are unable to process the majority of plants into something useful, so if you ever see someone laying on the ground eating grass, they are more likely to be taking ecstasy than knowing something you don't.

An array of microbes in the cow's digestive tract help break down the plant material, and methane is produced as a by-product of the very interesting process, which must then be emitted to prevent this gas from doing harm to the cows. A single cow approximately produces an astonishing 600 litres of methane a day, which, when multiplied by 1.5 billion, starts to become rather scary numbers.

The humble cow, as much a companion to humanity as dogs or horses, is descended from the Aurochs, which if you want a grammatical headscratcher is pluralized as… Aurochs. Larger than the cows they eventually evolved into as a result of domestication, these beasts wandered much of the world's surface for the past two million years, much the same as our good selves you may recall, and therein lies the problem. Thousands of years ago, they were popular in the Roman arenas as a battle beast, with a formidable temper and great horns to match.

As our need for farming land grew with our wildly uncontrolled population explosion, their habitat receded and receded. The last of the Aurochs breed died in 1627 in the Jaktorów Forest, Poland. of natural causes funnily enough, which was an ironic end for a species that had been hunted to extinction. The cows that descended from them of course lived on, and as our population continues to grow unabated, our need for burgers and steaks in much of the world is only set to increase.

Can the atmosphere be fixed, however? We have a lot of clever people in the human race after all, amidst the reality TV watching detritus that makes up a significant proportion of the remainder.

Methane is easy, we'd have to switch our diets away from meat and use the land to grow vegetables instead, a perfectly reasonable, varied and healthy diet can be achieved using a fraction of the land area, water and other resources that are required for the cows, pigs and chickens. Is that going to happen? More on that in the next chapter, but spoiler alert, I seriously doubt it.

Getting off carbon is harder, but still not impossible. We see and hear about efforts to become carbon neutral all the time, and slowly, painfully, the human race generally seems to be staggering in the right direction.

The problem however brings me back onto my favourite subject, the Tipping Point. It's not about what we do from this point on, it is what damage has been done already.

If we stopped all harmful emissions today, right this very minute, the planet will continue to warm for at least another hundred years. Ice caps are going to melt, people who live in coastal areas better start strapping on their water wings. Believe me, dear readers, this is non-negotiable. Our populist political leaders standing on their soap boxes (which won't be high enough to keep their feet dry) remind me often of King Canute, who famously (and apocryphally as it happens) ordered the tide out, to no observable effect. Of course, this didn't actually happen, as is the case with many of the best stories. Canute incidentally was King of England, Denmark and Norway all at once in 1028, but his legacy is ill-remembered nowadays thanks to the Norman invasion of England in 1066 just a few short kings later (by way of Harthacnut, Edward the Confessor and Harold the second) completely changed the geopolitical landscape. Northern Europe might be very different now if the Viking / Anglo-Saxon roots had prospered instead of the Normans.

Even now, a disturbingly short amount of googling will find you on the websites of naysayers who insist that climate change and the greenhouse effect are not a real thing, they are somehow leftist fabrications designed to keep good honest capitalists down. The most oft-mistaken argument for the non-existence of climate change is the delay in air temperature. Science has an easy answer to this, not that the capitalist corporations driving the inhabitants of Earth towards Doomsday care to listen. Free Carbon in the atmosphere takes more than 40 years to reach its full potential, so the temperature increases we observe now (in 2019 at the time of writing) are based on the greenhouse emissions from 1979. Even then things were getting bad, but we definitely hadn't got into the swing of things. By the time my nieces and nephews are into their forties, the amount of carbon we are busy emitting today will bite and take hold.

As in any complex system, ecological or otherwise, there are always feedback loops to contend with. The ocean absorbs some of the $CO_2$ we emit, but the temperature of the water impacts how much can be absorbed. The more we heat the water, the less $CO_2$ it will take on our behalf. On and on it will go, further exacerbating the situation.

And we're just getting started here people. Deforestation removes the ability for plants to come to our aid, all in the name of making another few quick bucks, and clearing some space for yet more farmland, or houses for our spiralling population, or industrial estates to deliver them the products they have been programmed to crave. I've already talked about the melting ice caps, this is another runaway effect that will see the sea level rise by two meters in the next eighty years, displacing a good two hundred million coastal inhabitants who will migrate inland, and further consuming our remaining resources and hampering our ability to do anything about the problems.

The melting of the ice caps will change the albedo (reflectivity) of Earth, reducing our ability to reflect some of the Sun's warming heat. Result? More temperature increases.

Faster and faster we go. The metaphorical runaway train containing everyone on our poor abused planet picks up more and more speed, the ravine is ahead of us and the bridge is out. People are still trying to save humanity, people like Russ George, but what nobody wants to admit is that it's too late. And why would they, these are people with husbands, wives, sons and daughters. They play badminton on a Thursday night, have friends over for dinner parties. It's easy to see why nobody wants to talk about this, it really is.

While attending, or at least passing by a recent protest of the awesome 'Extinction Rebellion' in London, I was struck by one of their slogans: 'The planet can do without us, but we cannot do without the planet'. 100 percent true. I don't even need to debate something so fundamentally accurate. The problem of course with Extinction Rebellion is that the powers that be, those who make these big decisions on our behalf (often old white men) don't care, and more importantly won't care. Their way of life, and their vast profits to be gained from the rape of the natural environment are yet to be curtailed, and nor will they while the corporations own the politicians, something which won't be changing any time soon, if ever.

Power to the people. Honestly, if there was a more pointlessly ineffective statement then I don't know what it is, but I don't watch Reality TV, so I appreciate I am shielded from the lowest percentile of banal utterings.

Even if it were technically possible, how can we reform our planet when the corporations who so stand to short term benefit from destroying our long-term future hold sway over our political landscape with deep pocketed lobbyists and campaigners. The short answer is there is no way. A major blocker is increasing economic regulation without a corresponding reduction in the human quality of life. My response to this would be short and disdainful. Of course, we're going to have to reduce quality of life, we're also going to have to reduce the number of people, full stop. We're overpopulated, living in a receding ecosphere and running short on resources, especially energy. The maths is so unbelievably, ridiculously, inescapably simple. The governments of the world couldn't reduce the people's quality of life without the people rioting, and ultimately removing them from power. It's political suicide, and so the runaway train continues onwards.

And even an Aurochs knows when his or her time is up.

# CHAPTER 7 - MEMOIRS OF GREEN

First and foremost, I am of course trying not to get sued by Vangelis for his excellent track on the Blade Runner (1982) soundtrack, though that is by no means my only reason for naming this particular chapter. Looking back, it's surprisingly impressive how many of my ventures and decisions revolve around trying to avoid litigation of some form or another. It's not that I'm seeking out risk, but more I hope of an indicator that the old phrase about how dull it would be to go through life without offending anyone stands tried and true.

And what's on the menu this chapter I imagine you asking, possibly with a certain slump of shoulder and roll of eye? Vegetables, I suppose would be the factually accurate answer, but more specifically some observations (with en suite sarcasm naturally) on the subject of how green the areas of our planet that aren't underwater used to be.

For this one, to start the story we need to go back further than you might think. 200 years? Nope. A thousand years? C'mon, you're not even trying. Cast your mind back to a few hundred million BC. In human terms it's a fair amount of time I'll grant you, but barely a leap year for the already adult Earth which was missing its carefree teens and probably well into lamenting how boring being an adult and having responsibilities is.

To measure it another way, the world had just embarked (in the grand scheme of things) on the Carboniferous period. Interesting word that, Carboniferous. It comes from the Latin carbō meaning coal, and ferō meaning 'I carry'. How about that – an entire geological era of the planet named 'I carry coal'. Can you guess what happened yet? At the start of this period, much of the land surface of the world was covered in lush, verdant rainforests. Our amphibian ancestors of the time still hadn't quite got out of the habit of hanging out in the water a lot of the time, I have a niece who is still very much of that opinion.

In the middle of that idyllic time, something changed. The climate. That's right folks, human beings aren't the only things that can completely cock up an otherwise pristine and tranquil environment, nature is pretty good at it as well from time to time. The exact cause isn't thoroughly proven, but volcanic activity is a strong contender, along with aridification (the land getting drier), although if that was the case the cause of the drought isn't understood either. Whatever the reason, much of the biosphere of Earth at the time died screaming. Well not actually screaming, not unless you believe a questionable M. Knight Shyamalan flick which was a terrible mistake for Mr Wahlberg.

Carboniferous. I carry coal. Honestly.

This 'minor' extinction event (minor in that it didn't rate the top five previously covered) was known as the Carboniferous Rainforest Collapse. Without the multitude of living plants, the concentration of carbon dioxide in the atmosphere crashed to one of the lowest levels it would ever be in the history and likely future of the planet. And what of all that dead plant material, I hear your cogs turning. Well that eventually became much of the coal we use today to inflict irredeemable harm on our present biosphere. What goes around, comes around, as they say – there is a dreadful symmetry in using the deathly ruins of almost an entire preceding ecosystem to precipitate our own demise.

I'm slightly cheating of course, we're meant to be focussing on manmade activity, but we have to start somewhere, and if you know anything about me by now it's that I am king of the digressers.

Let's get ourselves back on track, and fast forward back to nearer the present. As recently as 8,000 years ago Earth was still unspoiled, sure, she had a good few kilometres on the clock, but it really didn't show. A few million humans hustled and bustled about its surface, going about their primitive business in a way that did little to no harm to the ecosystem they inhabited. Life would no doubt have been hard, but it would also have been, well, decent, though humans are terribly good at failing to appreciate their situations until they have changed for the worse.

Then, around three thousand years ago, along came the Romans. Along marched I should say, no doubt in a phalanx of some sort. Rome itself, according to legend was founded by twin brothers Romulus and Remus on the banks of the river Tiber. After a number of adventures including being raised by a she-wolf, Romulus eventually killed Remus apparently over naming rights to the city, and so Rome was born in blood. Since their father was allegedly Mars, the god of war who raped their mother, I suggest you treat this grisly origin story with the same healthy scepticism as all the others.

The Romans built everything out of wood. You name it, they made houses out of it, ships out of it, and where they weren't making things out of it, they were burning it to keep warm and provide fuel for their forges to make their weapons of war. Sometimes though it wasn't even the trees they needed; it was the land they stood upon. At its height the Roman empire consisted of more than 50 million people, with a million of those living in Rome itself. For its time, it truly was an impressively extravagant empire, but with that brought an awful lot of mouths to feed. Extensive agricultural infrastructure was the cornerstone of the empire, and more and more land was sequestered to this purpose as the population continued to expand. The majestic untouched forests that at the time grew throughout Europe and the Mediterranean region were systematically chopped down in order to provide the wood the empire needed, and the space for agriculture and living. As the empire continued to prosper, the population grew, which needed more towns, more space, more agriculture, more wood, all of which led to more population, which needed more space, more agriculture, more wood. And that is why we need Eddie Van Halen on guitar.

As the Romans consumed more and more of the natural resources, the empire surged outwards in all directions in search of new untouched wilderness to appropriate, which brought them into violent contact with other tribes across what is now Europe, Africa and Asia. A large standing army of more than half a million men (no women, not a very progressive time unfortunately) facilitated the militaristic expansion. Farmers were drafted in to bolster the army size, leaving less people to tend the land, putting additional pressure on the expansion efforts. The Roman army didn't like forests because of their tendency to be filled with savages (another word for 'not Roman'), so wherever the army went, the forests were systematically removed to reduce the risks of an ambush. I wouldn't have wanted to be a tree in the midst of the Roman empire, I can tell you.

So now you have the definitive answer to the age-old Python-esque question, 'What have the Romans ever done for us?' They chopped down so many trees it actually increased the amount of $CO_2$ in the atmosphere by removing such a significant proportion of the ecosystems natural carbon / oxygen regulation system, that's what. This wholly unsustainable empire finally collapsed after 1500 years of inertia, falling to political instability, foreign invaders nipping at their borders and endless religious squabbles. Honestly, anyone who looks at the Romans and thinks they represented some kind of high point of civilisation, serenity and order (along with all the incest and hereditary insanity) for the human species needs to seriously think again.

The deforestation didn't stop there of course, in fact it had barely got started. From around the 11th Century onwards, right after we'd struggled out of the dark ages, human population began to seriously climb, and with it, our need for land that didn't have giant tall plants taking up so much room would only increase.

The first tree (or at least the earliest discovered) is thought to have been the Wattieza, a type of fern completely unrelated to the fern trees that currently adorn rainforests and your nan's conservatory. As covered in earlier chapters, whole branches of evolution were wiped out by the periodic extinction events that befall this world (the Carboniferous Rainforest Collapse in the case of the Wattieze tree), leaving familiar looking forms to subsequently re-evolve again from the surviving ancestors. I find it amazing that organisms separated by millions of years can evolve in similar ways when subjected to the same stimulus. Perhaps though there is more order to the universe than I give it credit for, when considered on a long enough timescale at any rate.

Among the many unforeseen consequences of deforestation is soil erosion. It turns out, perhaps unsurprisingly that the trees provide something of a windbreak for the delicate surface, and more specifically, the loose, life giving layer on the top that some call earth (not to be confused with the name of a planet you may have heard me throw around), or even more simply, dirt. Soil is a fascinatingly complex mixture of organic matter (carbon-based compounds), minerals, liquids, gases and organisms that support everything on this planet that we know of life. Having started reading up on it, it's clear that it deserved a book all of its own, which of course has already deservedly been written many times over. We owe our very existence to the soil, it's one of the many obscurely shaped puzzle pieces that has enabled life to thrive outside of the oceans.

Without healthy soil containing compounds such as water and nitrates, crops become next to impossible to grow. We need crops to survive as a species, one only has to look at some parts of Africa over the decades to see what can happen to a country when crops fail; instability, poverty and finally anarchy. Even the most hardcore carnivorous meat-eater, who lives on an albeit terrible diet of animal flesh can't deny that the creatures they so enjoy consuming, grew up big and strong and ready to be consumed, on a diet that largely grew out of the soil.

No trees, and the soil tends to blow away. It's that simple and is one of the many unforeseen consequences of humanities thoughtless and aggressive expansion around the globe. It doesn't happen overnight, in some cases it can take centuries, and therein once again lies the problem. Human beings are unable to think about next week, let alone next year, generation, next century. It's always someone else's problem. But the day has finally come when it's our problem, and no matter how loudly the scientists shout and point angrily to their charts, the corporations and the governments still aren't going to lift a finger to do anything about it.

At the time of writing the UK is still in the grip of Brexit, possibly the greatest political flustercluck ever to hit this Sceptre'd isle, as Shakespeare once lovingly, if obtusely (as was his wont) described it. Again, could fill a book etc, but it would be considerably more depressing than one about soil, in case a budding writer out there is struggling to choose between the topics. In any case on a personal level, I have found the research and summation of humanities foibles in this book a considerable comfort to the prospect of an ill-advised, poorly conceived, politically motivated departure from the European Union. While I, like many other rational human beings wanted to throw open the windows and shout abuse on the twenty third of June, 2016, I now see, and perhaps you also start to, that such a farcical satire has absolutely nothing on the greater catastrophe that is taking place right under our noses.

Soil erosion is understood to be one of the major causative factors behind the collapse of the civilisation of Easter Island, which I wrote about in my last book and I believe it to be an excellent microcosm of what our species is undergoing now. I shan't repeat myself here save to highlight that a once prosperous tribe inhabited a beautiful island. Because of the fashion for the ruling class to have large amounts of clear space for their estates, lots of the trees were cut down. Without the trees the soil largely blew away, without the soil the crops couldn't grow, and without new trees they couldn't build boats to evacuate the population on, as is humanities style once it finishes up consuming the resources in an area. By the time the first colonial visitors arrived, the population had crashed, and cannibalism was in full swing. Another fifty years and I doubt anyone would have been left at all, just some funky statues to show for more than seven hundred years of habitation.

The soil that blows away doesn't just magically disappear either. Because humanity likes to live near water sources, that's where the majority of trees were chopped down. The soil that blew away went into the water, which becomes polluted, sludgy and unusable. Worse still, it has happened through history so often that a word exists for it, siltation. As far back in time as Ancient Greece the expansion of humans has caused siltation, which is a fancy word for ruining the water. It got so bad in places that the accumulation rendered harbours and ports completely unusable. The coastal cities of Ancient Greek cities of Ephesus, Priene and Miletus had to abandon their harbours because of the accumulated silt. Same story in Ancient Syria.

The Belgian city of Bruges, which as you will all know, is a 'fucking fairy-tale' has been around in one form or another since the Bronze Age but has existed in a recognizable form since the 9th Century. Some of the buildings still standing today actually date from as far back as the 12th Century, which bearing in mind the terrible conflicts that have periodically engulfed mainland Europe is really something. From the 12th to the 15th centuries, Bruges experienced its Golden Age, and was a place of culture and commerce known throughout Europe. This all changed however with the silting of the port, which forced the merchant princes to relocate their logistical empires 60 kilometres up the coast to Antwerp, which today remains among the largest ports in Europe. Sorry Bruges.

This is a classic example of what is known as a 'Progress Trap', where a town or city with an expanding population manages to bring about its own downfall. This first happened millennia ago in Asia Minor (now Western Turkey), and many times since around the world since then.

It goes a little something like this.

A site would be chosen for a city. More accurately, a tribe who had found a nice location – near ready sources of wood and water stopped moving around. With these resources on hand, they would grow faster than other tribes and over decades or even centuries, absorb those populations who couldn't stand the grass being greener over there. Wood is chopped down to fuel the society and make space for increased agriculture, which grows in population and sophistication. It wasn't just for fuel either, the ever-improving materials science of metallurgy needed colossal amounts of wood to run the forges and smithies of the time producing metals for defence, offence, and other tokens used in everyday life, in one of the earliest examples of consumerism.

All that lovely unprocessed wood happily growing as trees, for centuries in many cases. No consideration was given to replanting what had been cut down, and even if it had, it's likely the need for wood would still have outstripped the pace of regrowth, uncontrolled human expansion being what it is. When the local wood starts to run out, the woodcutters have to travel further and further away to cut down trees. The transportation of the wood equally becomes more and more difficult. Costs go up. The industries that need the wood wane and collapse with the cost of wood becoming too high to be economically viable. The smart ones see the difficulty in time and relocate to other cities. The less smart ones do not. The city enters a terminal decline and is eventually abandoned.

Centuries pass. The trees regrow in the area, and an enterprising tribe of humans founds a city. The words of George Santayana (and not Churchill as is often stated) 'Those who do not learn history are doomed to repeat it' have never been so true when we consider the Progress Trap. All the fault of human beings, reproducing exactly like a virus, I hear Agent Smith saying in the lovely languid tones of the superb Hugo Weaving in The Matrix.

While we're talking about cutting down trees, I'd like to make a special shout-out to Donald Rusk Currey, who in 1964 as a graduate student of the University of North Carolina cut down the oldest tree in the world. Prometheus was a bristlecone pine tree, named by naturalists who visited White Pine County, Eastern Nevada and were struck by the natural beauty of the region. They were definitely naturalists and not naturists incidentally – to my knowledge they were all fully dressed at the time.

Ironically, the stated goal of his scientific pursuit was to prove that some of the trees in the area were old, and in that he most certainly succeeded. To give him his dues, he had tried on four separate occasions to obtain horizonal cores from the tree before resorting to felling the old tree, and on similar trees in the region was able to successfully determine an age of around the 3,000-year mark. Nobody could have been more surprised than Donald, when after felling it on 6th August 1964 he found that the Prometheus' age was at least 5,000 years old, older than any other non-cloned organism on the planet. Timber. Incidentally the next oldest tree, Methuselah (these old trees have great names don't they!) at 4850 years young is also a bristlecone pine, clearly the species is an enduringly successful product of natural evolution.

Elsewhere on the African savannah, we are bringing down the ancient trees with more indirect, but no less deadly means. The distinctive and majestic Baobab tree live for hundreds and in some cases thousands of years, and across Southern Africa they have been dying in their thousands, and more specifically, the deceased tend to be the oldest and largest ones.

At first, an epidemic was suggested, like the Dutch Elm Disease that has ravaged the worlds trees again and again over the past century. Originally hailing from Africa, the name Dutch Elm Disease relates to the nationality of the two phytopathologists (a fancy name for plant disease specialists) who first identified the disease in 1921. However the cause in this case is more sinister. The team leader, Dr Adrian Patrut of Babes-Bolyai University, Romania had this to say, 'We suspect that the demise of monumental baobabs may be associated at least in part with significant modifications of climate conditions that affect southern Africa'. Put more simply, the warmer temperatures and drier conditions associated with our changing climate are killing off the most vulnerable organisms first.

Recent research (as of mid-2019) believes that more than 600 plant species have gone extinct in the past 250 years, twice the number of animal species that have met the same fate across that time. Something tells me that this is the tip of a much larger iceberg, but short of grabbing a magnifying glass and going out to count woodlice, I have to take the facts at face value. It makes sense incidentally that the animals disappear at a slower rate, since the animals have slightly more capability to adapt by virtue of being able to move to new areas when they surroundings become more inhospitable.

It's not just the loss of trees with their $CO_2$ absorption abilities that causes us great difficulty, it's what we do with the land after we've removed the trees which is as much part of the problem. We chop down the trees because we want the land, which gives us space to live on, and to grow food on. The smart money for food generation is on vegetables, the significantly less smart money is on the creation of meat to feed our age-old obsession with consuming the flesh of everything below us on the food chain.

Growing plants consume vital elements such as nitrates out of the soil. These elements are replenished over time by natural processes, but over-use of land for growing plants, either for our own consumption or for the animals we then consume, leads to the soil to all effects becoming lifeless and barren. This lifeless and barren land ends up going by another name: desert, and the process by which this comes about is imaginatively titled desertification.

I'm personally probably about 70 percent vegan now. It might not sound like much but considering I would have been less than 10 percent vegan just a few years ago, it feels like progress. One of the key things for humanities survival in anything like its current shape and size would be to quit the meat. I realise that is a controversial viewpoint, and especially hypocritical given I am a regular attendee at the annual Meatopia event in Tobacco Docks, London, but all I can say is that it comes under the remaining 40 percent and I promise I'll stop going at some point.

The ray of sunshine, both for me going to Meatopia, and humanity in general (and how rarely are those aligned) is the rise of the vegan burger. Pioneered by companies like Impossible Foods and Beyond Meats, these completely meat free burgers are entirely plant based, and in an amazing testament to the spirit of human ingenuity, taste exactly like the real thing. Now you might think that such a development is indicative of the hipster urban landscape we find ourselves uncomfortably inhabiting, but I was fortunate enough to try an Impossible™ Burger in San Francisco some months ago, and quite frankly, I was astonished.

Before setting out to San Fran, I'd set myself a mental post-it note to go visit a restaurant that served them and then as normal, the post-it note had promptly fallen off into the dark recesses of my mind. I was therefore as surprised as anyone to find myself looking at an Impossible™ Burger on the menu of the eatery I'd managed to wander into. One thing I know however, is if you want to get on and enjoy your life the way you are supposed to, you have to learn to listen to the universe when it speaks to you. This was just such an occasion. The burger was succulent, delicious, healthy and very hard to tell from the real thing. The research and development that must have gone into producing these is phenomenal, not to mention the ability to mass produce them and be resilient enough to be flung around the modern high-pressure restaurant kitchen.

One thing is clear, this is not just some fad. Don't believe me? As of mid-2019, Burger King are currently rolling out the Impossible™ Whopper across America. I firmly believe we are staring down the barrel of an end to our species' addiction to meat, and I for one could not be happier. When eating just vegetables we remain healthier, live longer, and treat the ecosystem we live in with a good deal more respect. There is more breaking of wind to contend with, but for those of us who can see humour everywhere, this is no problem at all.

Naturally, wherever you see a positive force for change, you see an equal and opposite lobby seeking to shut it down, kill any innovation and preserve the status quo. It was true for electric cars, it was true for solar panels and it's just as true for this new generation of meat-free products currently tempting us away from beef, pork and chicken. The cattle industry worldwide is a multi-billion-dollar industry, and a few people with little to no morals stand to lose an awful lot from a lot of people with strong ethics choosing what to put into their bodies. How dare they, after all. The result is the usual dirty tricks campaign, where obfuscation and misinformation are rife, the latest salvo of which at the time of writing is attempting to prohibit the use of the word meat when saying that the product is 'meat free'. Honestly.

Meanwhile, in the temperate parts of the world, rainforests, the vast, lush and humid stretches of land that play such an important role in the carbon / oxygen regulation of Earth continue to decline. Where once, a mere 8,000 years ago they would have covered 14 percent of Earths' surface (a little over 70 million square kilometres worth in fact), today they cover just 6 percent, with the remainder expected to be gone by 2060. The reason for this decline is wholly manmade, the demand for wood and open ground by our species has never been higher, 1.5 acres of it disappear every second, chopped down deliberately, or dying back due to the polluted environment. Stop and think about that for a bit, but don't take too many seconds, will you.

If you take a single conclusion from my diatribe, let it be this. Our entire ecosystem is changing around us, too slowly for the human eye to comprehend for sure, but it changes just the same. That change can be best described as 'from a habitat we can exist in, where we breathe, eat, drink, work, rest and play, to one where we do not'.

Go take a look out the window for a minute, enjoy that greenery while it remains. I'll be waiting for you in the next chapter.

# CHAPTER 8 – POISONING THE WELL

It's time to dive back into the oceans. Be sure to wear a wetsuit, as it's certainly not going to be any cleaner than the last time we explored it. Although we are by no means the only creatures guilty of this particular fallacy, our long and epic history of polluting the oceans started, as it always seems to, with pooping.

I'm serious. If there was a stream, river or lake, our distant ancestors seem to have been busy pooping in it. There is doubtless a psychological aspect, running water handily washes all our bodily sins away, out of sight and out of mind as it were. That is of course very convenient on the local level, but everything has to go somewhere. I suggest you get used to the phrase 'closed system', as it's extremely relevant to the predicament our species is busy sticking its fingers in its ears and humming noisily over the top of.

You don't need me to tell you this, you instinctively know poop has a load of bad stuff in it, so I forgive you if you want to skip this paragraph. A UNICEF report from 2008 stated that a mere one gram of human poop contained ten million viruses, one million bacteria, 1,000 parasite cysts and 100 parasite eggs. These days, the majority of it is treated in some fashion before flowing it into the sea, but that is nowhere near 100 percent effective and not everywhere has access to treatment facilities, and so into the sea it all still flows.

Moving swiftly on before everyone (including me writing it I assure you) puts the book down in disgust; let us now move onto more unsavoury but less graphically unpalatable aspects of the issue.

Everything started seriously going (flowing?) downhill of course when the Industrial Revolution swept the planet, as always casting a long shadow over the cesspool our Earth has slowly become. While the human population of this spherical ecosystem we like to call home had been steadily growing since the 1500s (with all the associated pooping that brought with it), things really took a turn for the worse in every way except short term human comfort in the 18th Century.

With the advent of the Industrial Revolution, greedy little men (I exclude women only because of the patriarchal nature of the time, and honestly my sisters, take the win of dissociation) all over the world decided it was cheaper and easier to pour their chemical waste from their grubby factories directly into rivers and streams rather than dispose of it properly. The same psychology throughout history was prevalent: If I put what I don't want into moving water it ceases to be my problem and becomes someone else's far, far away. I find this especially perverse when you consider how many of us find staring at moving water to be an incredibly calming and cathartically soothing experience; try not to think who upstream is trying to get rid of a few metaphorical (and sometimes literal) bodies, or just having a really satisfying wee.

It got so bad that by the 1960s, there had been several documented reports of waterways actually catching fire, including Cuyahoga river and Lake Erie in Ohio, from all the awful sludge that had been poured into them from the surrounding industries. And all in the time-honoured tradition of saving a couple of bucks on safe and proper disposal.

While industrial disposal has in many ways cleaned up its act (only when forced to by law of course) since those bleak days, major incidents still occur all the time in the developing world and occasionally even in places that really should know better. Another UNESCO report cites: 'Up to 500 million tons of heavy metals, solvents and toxic sludge slip into the global water supply every year'. Now even those of us who didn't pass maths at school can agree that is a lot of poisonous crap to be dumping into the worlds water each year. The word 'slip' also has a rather accidental feel to it that doesn't quite imply the right amount of criminal negligence or incompetence verging on a war crime against our own species.

Sadly, nary a one of our shabby corporate CEOs and board members over the past few generations has ever uttered the words 'Wait a second, Earth is a closed system, we shouldn't be dumping our toxic by-products without securing them properly for future generations'. There are exceptions of course, honour among thieves and all that, but we're here to examine trends, not aberrant outliers, no matter how noble they might very occasionally end up being.

The terrifyingly rapid industrialisation of China over the past 50 years is a perfect case in point. Greenpeace recently reported that approximately 70 percent of China's lakes and rivers are now polluted with industrial waste. Notwithstanding the choking marine life, spare a moment for all those poor people crammed like sardines into their overwhelmed cities; with no choice but to keep on drinking the water provided for them by the very same government that has facilitated the pollution in the first place.

An indirect yet nightmarish impact of all the toxic waste worthy of our perusal is what the scientists call 'Harmful Algal Blooms', more sinisterly known as 'Red Tides'. It turns out, the kinds of by-products our less scrupulous factory owners over the decades have tossed away are absolutely delicious to some of the normally harmless denizens of the oceans, namely certain varieties of algae. In terms of their impact, as you may have gathered from use of the word 'harmful' in the name, it's not going to be a good.

Algae are a collection of harmless microorganisms that have the ability to conduct photosynthesis. In that regard they are a close relation to our ultimate ancestor, certainly closer than we are at any rate. Examples of algae are seaweeds such as kelp and plankton, not to mention the pond scum that middle aged suburbanites are forever scooping out of their gardens to try and protect the poor tadpoles trying to make a living for themselves.

Amongst other things, algae feed on the tiny amounts of phosphate and nitrates floating around in the low parts per million. The issue hits when they suddenly get a huge influx of such chemicals, courtesy of our unscrupulous factory owners. Much like the single celled life form we owe our entire existence to (she should probably get partial credit for the book really), certain chemicals make them multiply like crazy. While this doesn't sound harmful in itself, consider that they also consume oxygen out of the water to fuel their growth spurts. At normal levels of growth, the consumed oxygen is easily replaced out of the atmosphere, all part of nature's majestic balance, but acting under the influence of these chemicals, an imbalance forms, removing significant amounts of dissolved oxygen out of the water.

Less dissolved oxygen in the water is an extremely bad thing of course for our aquatic brothers and sisters from other species, dependency on said oxygen to breathe being another trait we have in common with all of them. When and where a bloom occurs, aquatic life tends to die in its footsteps (fin-steps?), choking to death in water that is no longer breathable for them.

And to cap it all off, a by-product of the Red Tide is a very nasty waste chemical called 'Domoic Acid', which seeps through every part of the food chain, and should absolutely not be considered one of your five a day – in fact it's been known to cause brain damage, seizures and death. As recently as 2015, fisheries in Washington, California and Oregon were closed due to toxic levels of this acid in shellfish, so this problem is most definitely not a relic from a bygone age.

Elsewhere in the deep oceans, something even more cataclysmic is taking place. A study published in January 2018 by marine ecologist Denise Breitburg found that the ocean has lost seventy-seven billion tons of dissolved oxygen in the past 50 years, almost 2 percent of its total concentration. While similar to the Red Tide in its effect, albeit on a monstrously larger scale, this blight that is killing off marine life has a difference cause to the algal blooms we've examined, something that is all too familiar.

Us.

It comes down to physics (doesn't everything). Don't worry, if you failed physics at school or had a hard day and don't want to try and remember it, I'm going to explain why (stop groaning, it won't be that bad).

Consider first the simple fact that warm water can't hold as much dissolved oxygen as cold water. This is a basic tenet of chemistry, and as such isn't really up for debate, though as any creationist will tell you, that doesn't mean anything. Despite that, it is commonly accepted now that globally, sea temperatures are on the rise thanks to Climate Change, humanities centuries-old club we use to bludgeon our poor mother Earth with while simultaneously hiding the club behind our back.

Warm water, like warm air rises, while the cold sinks, all of which means that due to Climate Change, oceans are being covered in a shallow layer of warmer water. The number one influencer on the temperature of the oceans is not the underwater volcanoes and oceanic vents that are scattered through the deep chasms in the middle of the oceans, no, you may be surprised to know that it is the atmosphere. It makes sense when you think about it, there are approximately 36 million square kilometres of oceanic surface area, the entirety of which kisses the atmosphere, and has done for billions of years. What happens in one is bound to affect the other in short order.

So, when we affect the delicate balance of the atmosphere by pumping it full of greenhouse gases like $CO_2$, we increase the temperature of the atmosphere, which then increases the temperature of the water. That might be useful to you if you're planning a swim and wish to avoid hypothermia in front of your loved ones while on a family break, but for the oceans, that increased temperature it makes oxygen absorption from the atmosphere that much more difficult, further exacerbating the issue. All of which demonstrates in one more decisive way, if the ecosystem was a train, its uncontrolled momentum towards a grim and uncertain end would absolutely qualify as 'runaway'.

This research surprised even one as jaded and cynical as my good self (or would that be chaotic neutral self?). After decades of our sadistic, if largely oblivious torturing of our planet, we really have done it this time. How has this not become the number one news story all around the world, I asked myself? After all, the news services love a bad news story the way I love sweetcorn fritters, they just can't get enough and make me sick.

The only conclusion I could come to was that this story wasn't interesting enough as it wasn't hurting people on a personal enough level to be appreciated, which is a depressing realisation about one's species, even while knowing deep down it was true. Out of sight, out of mind may well end up being our species' epitaph. Perhaps the news services of the world knew, and rationalised it away - after all, it's not like much can be done at this point, as the effect is probably unstoppable, particularly when you have questionable individuals in positions of power who won't even admit the problem exists let alone plan to take action. And who can blame them really, they might after all make less profit if they took such drastic action as to notice and act upon our dying ecosphere, especially when it's always (so far) been a problem for the next generation.

But of course, the truth is they didn't rationalise it away at all, they either didn't know or didn't care, and I know where I would put my money, if it wasn't all in Bitcoin.

How such people in power, who may have an inkling of what is to come next inevitably seem to have produced children they appear to care about, is utterly beyond me. In the first book I talked about an old friend of mine who once remarked about his children 'it all changes when you look at them for the first time'. At the time I likened this dramatic personality shift caused by our Wiring to Invasion of the Body Snatchers, which might have been amusing, if only to me, however I failed to look at those of us it doesn't seem to affect. The rich and powerful. Those who should have the worldliness and astuteness to look around at the planet on fire, at our dwindling resources, at our polluted ecosystem and think, 'hang on a bloody minute!'. Yet it just doesn't happen. I can only equate this with the surgical removal of empathy that appears to accompany the ascendancy to the ranks of the 1 percent, the rich and powerful elite who, having stepped over everyone to attain their wealth and power, will do anything to avoid giving it back.

Another casualty of our war on water is the Coral Reefs (if you don't know how to pronounce Coral, look at how Sherriff Rick says his sons name on Walking Dead and you have it). In case you don't know (and feel free skip over if you do), coral is a type of invertebrate (sans backbone) marine organism. In that respect it is of a very old, let's say classic design, evolutionarily speaking, and like the algae that swirls around it, coral has much more in common with some of our earliest common ancestors than we do.

The reef part of a 'Coral Reef' is made up of calcium carbonate (limestone to its friends), a porous material actually secreted by the coral. I wasn't joking earlier (for once) when I talked earlier on about the endless diversity of the creatures that have evolved on this planet. Think about it for a moment, the reefs are one of the most beautiful types of undersea ecosystems to exist, and they are created by a lifeform with such ancient and wacky DNA that it's able to secrete a type of limestone as part of its internal bodily chemistry.

The largest and greatest of these that I'm sure you will all have heard of, is the Great Barrier Reef off the coast of north-eastern Australia. Not a single reef, it is in actual fact a complex system of almost three thousand individual reefs, and collectively they are home to more than 25 percent of all seagoing species that are known to exist. Talk about a popular neighbourhood. Suck it, Shoreditch.

It is also the largest structure on Earth ever created by living organisms and is visible from space. Suck it, Great Wall of China.

Enough sucking it, we must keep moving forwards. In 2016 during an unprecedented heatwave, more than a third of the corals comprising the Great Barrier Reef (an estimated five hundred million individual organisms) died over a nine-month period. Let that sink in for a moment. Yep, it's a lot. The cause was, you've guessed it, Climate Change, specifically in this case the change of climate from a cooler one where a load of these little critters were alive, to a hotter one where they weren't anymore.

I'm going to have a minute of silence here for the five hundred million creatures our collective hubris and greed exterminated before I continue. You take as long as you want, I won't time you.

The total body of water on Earth, also known as the hydrosphere, is estimated at over 500 billion cubic kilometres worth, which is of course a phenomenal volume. It first formed four billion years ago from the condensation of water vapor, which as we covered in earlier chapters, came from the cooling rocks that made up primeval Earth, and even more brought by the relentless impacts of comets, asteroids and meteors that hurtled around our young solar system. The most common of these meteors, of the carbonaceous chondrite variety, contain up to 20 percent water. Even unfolding in the slow motion of geological time, it would have been quite the sight to see all those impacts, each one changing the planet very slightly for the better, subjectively speaking.

The amount of water on Earth has not changed significantly since that time, what with this closed system we like to inhabit, from the deepest oceanic trench ten kilometres deep to the uppermost regions of the atmosphere ten kilometres above our heads. A small amount of water is lost to space each year through the aforementioned process of photodissociation, but not enough to have made a significant difference, even over the billions of years of the hydrosphere's existence.

With so much water literally at our fingertips, especially if you are among the 80 percent of the worlds human population who lives near to a coastline, it might surprise you to know that we are in significant danger in the very near future of running short of it. There will always be enough to go for a swim, of course, not that you'd want to with all the human crap (some of it literally) floating in it, but of the 70 percent of Earth's surface that is covered by water, just 2.5 percent of that is fresh, the remainder being heavily salted. Of that 2.5 percent, just one percent of that is easy to get to, with the rest locked up underground or in glaciers. All of which means that of those more than 500 billion cubic kilometres of refreshing, life-giving water (especially when frozen, cubed and served with rum), only 0.007 percent is available to us to drink and keep 7.5 billion people and rising every day alive.

As a side-quest, a mote of curiosity struck me here around why the oceans are salty, and equally why that makes it no good for us to drink it. After all, it seems a cruel quirk of fate that our species could reach a time in the near future when we run out of clean water, with so many of us living within sight of large bodies of water.

The salt in the oceans originates from the rocks on land. It turns out that rain is ever so slightly acidic, not so dangerous that we should legislate against pop videos and Kung Fu film climaxes from taking place in the rain, but still enough to make one pause. The acidity comes from dissolved $CO_2$ carried along with the rain (who put so much of that into the atmosphere I wonder – looking at you, humanity). This acidity erodes the rock over long timescales, and the rainwater carries the eroded waste products into the seas. These waste products include, among other things, sodium and chloride, which make up the majority of the salt content of the ocean. I read one fascinating article that said if the volume of salt in the oceans could be removed and stacked up on the land, it would cover Earth's surface to a height of 500 feet.

In Bad Wiring, I talked about why we can't imbibe saltwater, so I won't belabour the point here, suffice to say a person can die of thirst while drinking saltwater, due to the bodies inability to process water with a salt content greater than two percent. We nearly can, since the concentration of salt in seawater is 3.5 percent, but nearly doesn't save the lives of all those shipwrecked sailors through our history of traversing the oceans.

We can of course remove the salt from the water, a process known as desalination, but it is costly and energy inefficient. I was once on a boat sailing in the Caribbean, I'm not a rich man by any means, but a friend very kindly paid the yacht rental for a week and in return I had to work hard helping him crew, and wrangle his young family who wanted nothing more than to climb or fall into the water every five minutes. But there I was. On the back of the yacht there was a cylinder mounted, which extended down below the waterline. Being unfamiliar with it, I inquired after its function, and was amazed to discover that it was a portable desalinator. As the yacht carved its quiet, dignified course through the waters around Norman Island, on our way to a cocktail bar on a pirate ship I seem to recall, this technical marvel blew my mind. The forward motion of the boat was all the power the desalinator needed, as water was driven into the tube, going through a series of filters and ultimately topping off the water tanks of the boat, needed as my friends daughter was fond of very long showers and had already depleted the full water reserve once on the trip.

As amazing as that technology is, it remains applicable on a reasonably small scale only. The largest commercial desalination plant is Ras Al-Khair in Saudi Arabia, which produces a million cubic meters of clean water a day. At present, desalinated water is used as drinking water for approximately four percent of the planet's human inhabitants, the vast remainder rely on fresh water sources, which continue to be diverted into watering livestock, growing cotton and tobacco, all the things that we fundamentally don't really need. All the while, the population continues to go up and up, requiring more and more water to sustain. Necessity being the mother of invention, I fully expect the next few decades to see great strides forward in terms of desalination technology and efficiency. If you were looking for somewhere to invest your money as civilisation continues its inexorable decline, you couldn't go far wrong with desalination.

According to the United Nations, by 2025 there will be 1.8 billion people living with the threat of water scarcity. To quote the old saying, any society is three meals away from revolution, which makes me wonder how many glasses of water are we away from anarchy and barbarism? How thirsty does a child need to become before their desperate father considers caving someone's skull in for a bottle of mineral water they can take home (still of course – like all sane people I can't abide sparkling water).

Wholesale changes are needed in how water is collected, treated, conserved, used and distributed. As you will no doubt have gathered by now, wholesale changes are not exactly what our disparate, bickering species does best. Sure, the U.N. is going to pass a bunch of serious sounding resolutions, which will doubtless take years, if not decades of debate. I can't escape the mental image of those conference rooms in 750 United Nations Plaza, New York City, with all their delicious bottles of mineral water laid out for the delegates. Those resolutions are going to mean precisely nothing.

Meanwhile we continue to overwork the land with agricultural techniques that would make the early Sumerian farmers throw down their hoes in disgust.

Meanwhile we continue to operate enormous herds of animals so we can consume their flesh.

Meanwhile we obsess over short term consumerist clothing, buying shirts from fast fashion retailers that will literally have rotted off our backs before we get home.

Across the salted pond in the United Kingdom, the Environment Agency is equally concerned. Early in 2019, the agencies head, Sir James Bevan warned of an 'existential threat', highlighting that in 20-25 years' time there will not be enough water to meet demand in the UK. We may even experience rolling water blackouts (wet-outs?) as the available supply is rotated between areas in a desperate attempt to prevent a total breakdown of supply. One issue preventing the increase of supply and storage of water is the regional opposition to more reservoirs. A proposed reservoir near Abingdon in South Oxfordshire has been repeatedly blocked since it was first proposed in 1976, with protesters citing an increased risk of flooding in the area, and traffic disruption during its construction. Feel free to snort at the second reason, I know I did.

I mean, I totally get it though, nobody wants an instant, unlimited supply of fresh water literally right on their doorsteps when humanity enters a grim uncertain future of water shortages. We're not idiots, are we...

Sir James calls for us to reduce our water usage in a number of sensible ways including showering instead of bathing, making sure toilets, washing machines and dishwashers are water-efficient (i.e. not still using white goods that were a gift from mother in the 1970s).

All these measures though are stopgaps, and regional efforts in the closed system of Earth are largely pointless when great swathes of the rest of the world will continue to use and abuse the ecosystem with impunity.

All of which means, anything our species does to tackle upcoming clean water shortages will ultimately prove to be far too little, and far too late.

## Part 3 – End of the Line, My Valentine

*Each of us must work for his own improvement, and at the same
time share a general responsibility for all humanity.*
Marie Curie

## CHAPTER 9 – PATRIARCHAL PLANET

A couple of chapters ago I espoused the potentially controversial
opinion that we live in a patriarchal world that is in denial and took it
further to strongly imply that I think man specifically is at fault for our
present dire ecological circumstance. Not woman, man. Well, it's high
time I explained myself. I know what you're thinking, it's quite unlike
me to elaborate, especially at great and potentially tedious length.

Yes, that's right, I called it. Patriarchal. Given how many thousands
of years it's been since our humble beginnings scraping knuckles along
the floor and acting like a handprint on a cave wall was a Van Gogh,
it's frankly embarrassing how far we haven't come. It was a simpler
time of course, the women stayed at home, the men went out and did
all the work, and there was an uncomfortable amount of non-
consensual sex. That may sound a little like the 1970s, and you'd be
forgiven for thinking so, especially those of you who are of a certain
age and grew up in that toxic culture.

First off, I should make it absolutely, crystal clear that this is a bad
thing. In no way am I endorsing the current state of our society, I
consider it totally abhorrent, at best completely unfair to the 51.9
percent of the human population who possess XX instead of XY

chromosomes, and at worst tantamount to a crime of the highest order, almost without equal in our egregious history.

In selecting the topics I wanted to cover in the search for our particular Tipping Point, I wanted to include this section on our patriarchal society to shine a light on the hilarious-if-it-wasn't-so-fucking-sad denial by all those who would disagree with me. While I may not be able to put it right, at the very least I can call it out and say it for what it is, which if you squint a little might still feel like a hair's breadth of progress.

Believe me, if I could see a way to fix it I would do so, without hesitation, in fact this would actually be exempt from my usual self-educated 'pull up a deckchair to watch the end of the world' apathy about the future of our species –that's how massively unfair I consider it is. I even used italics, which shows you how serious I am because I almost never reach for the old Control-I.

Allow me to explain.

Looking first to our old friend the Oxford Dictionary, we get this definition to kick us off: 'Patriarchy is a social system in which men hold primary power and predominate in roles of political leadership, moral authority, social privilege and control of property'.

The urban dictionary however has this definition: 'The bogeyman that feminists blame for women's problems or under-achievements because their big-girl pants apparently don't fit'.

Do you see my point? The first is factually accurate (something the Oxford Dictionary is noted for I might add), and is therefore bad, assuming I convince you that it's the truth we all currently live under. After all, it is of course fundamentally and completely unfair in a world where both genders are equally intelligent and equally deserving.

The second is not even remotely factual, and plain sucks because it's hurtfully endemic of the prevalence of the sexist, misogynistic Neanderthals who still hold so much sway in our civilisation, especially over our main information transfer and communication mechanism, the internet.

This throwaway piece of sexist humour actually made me legitimately furious. For starters I think we can all agree that a male human wrote it. I can hardly see a female human typing that in with a self-satisfied nod and slight furrowing of the Neanderthal brow, were she even the greatest satirist the world has never seen.

Never has 'If you're not part of the solution you are part of the problem' been so true; the dickhead males who perpetuate such stereotypes have absolutely no business procreating and passing their pathetic excuse for wisdom on to the next generation. They should also have their internet privileges taken away, as clearly behaving like rational mature and respectful individuals is utterly beyond them.

Patriarchy literally means 'rule of the father' and comes from a Greek word meaning 'father of the race'. There is strong evidence that humanity didn't start out patriarchal, in the dim antiquities of pre-history, our hairy, cave-dwelling forerunners of the Palaeolithic period appeared to enjoy an egalitarian social structure, contrary to the popular opinions about club wielding Cro-Magnons dragging women in leopard print bikinis around by their hair.

I caught myself writing the word 'brethren' just there instead of forerunners, until I caught its obviously masculine connotations – brethren is an Anglo-Germanic word meaning fellow members of the same group, as a plural of brother. Hardly gender-neutral is it. Apparently 'Sistren' was also in use in the UK until the 16th century where it fell out of favour, coincidentally around the same time that 'Parliament' was formed in the UK. Parliament was touted as a great stride forward for society, a move to a true democracy away from a monarchy. Or to put it more specifically, a move away from a single old white man or woman having ultimate power in the country while lining their pockets, to a room full of hundreds of old white men who made all the decisions while lining their pockets instead. I apologise also for the gross oversimplification there, I neglected to mention the old white guy in charge of all the old white guys, who is even busier lining his pockets. Probably just a coincidence though, right?

Back to the Cro-Magnons anyway. Given we're talking about a time long before writing things down and passing them on became fashionable, what little evidence can be gleaned from archaeology and anthropology points to patriarchy becoming popular about the same time as agriculture and animal domestication and husbandry around eight to twelve thousand years ago. As our tribal ancestors grouped together into larger and larger tribes, it became clear for reasons of social order that someone should be in charge. That's where it all started to go wrong of course, because of course the men decided that it should be the men who made the decisions. Physical upper body strength is the key here. Women have unsurprisingly evolved differently to men, with certain child-rearing necessities that have diverted their evolution down a path away from brute strength being a key qualifier. And that's how it all started. If a man couldn't get what he wanted, or a woman disagreed with a decision, or disputed a claim to property (including her own body now I mention it), raising a clenched fist tended to resolve such disputes.

Rule of the fucking father, eh....

Some researchers theorise that the origins of patriarchy stem from the origin of fatherhood as a social responsibility around 6,000 years ago. Prior to this the tribal system held sway, the young were cared for within the tribe, the mother was easily identifiable at least in the early days from the evolved ability to feed the baby essential nutrients in her breast milk, but the father was long gone. He would have still been part of the tribe of course (this is sometime before slipping out to buy some smokes and not coming back became fashionable), but there was no emotional attachment, no familial connection from the male contributor to the genome. Other children in the tribe could be your brother or sister, but who would know? As the tribes settled down, transitioned from nomadic into communities, it became harder for the men to shirk their responsibilities. So goes the theory anyway.

Friedrich Engels, a famous Marxist believed that patriarchy came with advent of ownership of private property, something that even in those early days of communal society, men had always controlled – ownership of property would pass through the males in the family line while the females took care of the house and children. That being the case, it's another fine mess capitalism has gotten us into.

And so the trend continued. Around 5,000 years ago in Mesopotamia, Persia and Ancient Egypt we can see evidence of similar restrictions put upon women. Men did the important stuff like run an empire and fight each other, while women did the important stuff to run a home. The macro and the micro. However, the fact that totally passed those men by is that while you can have homes without an empire, you can't have an empire without the homes.

The Ancient Greek philosopher Aristotle, considered by many to be the Father of Western Philosophy in his writing referred to his belief that women had colder blood than men, which made them inferior in his eyes. Given the influence and importance his viewpoint has had on the world, he himself has a lot to answer for, and one has to also wonder about the rigour of his scientific experiments to form such a wildly incorrect theory. Also, I'd pay a lot of money to see him go a few rounds with Ronda Rousey, then see what he thinks about how inferior women are.

Ancient Egypt also deserves an honourable mention for being an exception to this ancient growing trend of marginalising women. Yes, there was an expectation that women had to bear the responsibility of nurturing the young, but they could also run businesses, own property and even become independently wealthy. There were female leaders in the religious caste, along with female practitioners of medicine. More than one of the pharaohs was of the female persuasion, including of course, Cleopatra, who has become an iconic figure in our cultural history. Disappointingly, Cleopatra is remembered mainly for the affairs she had with Julius Caesar and Marc Anthony rather than ruling for 21 turbulent years, which takes much more tenacity, intelligence and cunning than bedding an idiot with an erection. It wasn't to last though, in fact Egypt's progressive culture died with Cleopatra who was the last ruler of the Ptolemaic dynasty. After her passing on 12th August in the year of someone's lord 30 BC, Egypt reverted to a much more conservative, male dominated state, which along with pretty much the rest of the world, has continued up until the present day.

Despite this reversion, why didn't this catch on though, if it was so progressive, and so successful then slowly and painfully, surely other cultures should have been able to follow their lead? Well, it turns out, if you don't leave any written records behind, your teachings are hard to follow. We've all kicked ourselves for forgetting something we failed to write down, now imagine you're an entire empire. And I'm sorry, Hieroglyphs don't count if only a well-educated elite can read them and you hide the Rosetta Stone a little too well in a quiet backwater village in the middle of nowhere called... well, Rosetta.

Way out East a few hundred years earlier, Confucianism was wreaking similar havoc in China. Described depending on who you ask as a religion, a way of life, a tradition or a philosophy, it quickly became a core part of Chinese culture, and has been the official religion of China on more than one occasion. Confucianism states that a woman's value is closely associated with her loyalty and obedience, or to put it another way, subservience. Thanks a lot, Confucius. Such was the brainwashing that went on in that era, that a disciple of Confucius, a woman named Ban Zhao took the trouble of writing a book to highlight that a woman's primary goal is to find a patriarchal figure to subordinate themselves in front of, and that women were so fundamentally mediocre that it was best to not worry about such concepts like talent or intelligence. Ban Zhao was a product of her times, I like to think that in a more enlightened setting she wouldn't have prostrated herself quite so fully before the prevailing patriarchy. Even in more recent times in China, for example the Ming Dynasty (1368–1644), the Confucian values were alive and well. Widowed women were forbidden to remarry, and it was fine for women to not marry at all, but only so long as you remained chaste forever. I don't know about you, but I can barely manage three months without some kind of congress.

Over in Europe around the same time, philosophers were still nodding eagerly along with Aristotle. Sir Robert Filmer penned his epic 'Patriarcha' in the 1650s, where he espoused the divine right for kings to rule, since they apparently inherited that right all the way down from Adam himself. I'd love to read his citations for this rather wild claim. Fun fact here, there's a great legend that Adam was divorced when he met Eve. According to the Jewish Mythology that much of the Christian Bible is based off of, he was first partnered up with Lilith, but she left him after refusing to become subservient to him, and apparently became the consort of an archangel called Samael on her way out the door. It speaks volumes for perpetuated perceptions of female fidelity, and the hypocritical patriarchal reinforcement that embody religious mythology the world over for generation after generation to eagerly consume.

Things slowly started to take a turn for the better in the second half of the 18th century, as the world grew in sophistication and the views of the old world started to be challenged. Denis Diderot, founder and chief writer on Encyclopédie, the literary pinnacle of the so-called Age of Enlightenment wrote in the tome '...reason shows us that mothers have rights and authority equal to those of fathers; for the obligations imposed on children originate equally from the mother and the father, as both are equally responsible for bringing them into the world...'. Wise words, and a long overdue observation that it does indeed take two to tango.

The mortal enemy of the patriarchal society is feminism, the movement, or movements that advocates an equalisation of rights and privileges between sexes. The word feminism was first coined by Charles Fourier in 1837, though being French he said 'féminisme'. Fourier, a utopian scientist of the time (isn't that a lovely job title to be known for) was a strong supporter of women's rights, who he saw as equals, he even observed that men and women had a wide range of sexual preferences through their lives, which should be enjoyed provided everyone was a consenting grown up. However, for all his forward thinking, he was also an economic anti-Semite, believing that trade was one of societies great evils (no argument here), and that this was all the fault of the Jews (err... really?) who in his view of the future should work as some kind of slave-caste in the new world order. Still, as I said before everyone is a product of their time, and we have to take into account the social thinking that would have had an influence on him.

On that subject, I was recently amused by a tabloid exclusive on the subject of Albert Einstein being racist, based on some travel diaries he wrote in the early 1920s while travelling in China. Honestly, that has to have been a really slow news day for the unimaginative journalists that dredged that up to vent some of their unwell-earned righteous indignation.

Extra, extra, read all about it!! Person from 1920s has opinions common to the time he lived in!!

To be fair though, what else is the news outlets of the world going to find to report on, since they are steadfastly ignoring the hopeless situation that humanity has found itself in with a blithe determination that borders on the pathological.

Feminism has historically been divided into four waves. The first was the suffragette movements of the 19th century, where women decided it was high time they should have a vote, and some astute male politicians of the time realised that to support this effort would lead to an awful lot of votes coming their way. This was followed by the women's lib, or liberation movement of the 1960s and 1970s, where feminists campaigned for legal and social equality and according to their detractors, burned a lot of bras. In the early 90s a third movement covered the desire for greater individuality and diversity, began with the Riot Grrrl punk movement in Washington State. The most recent wave (and believe me I didn't use the word final very deliberately) covers the rise of the #MeToo movement, which was instigated by the actions of a vile Hollywood movie producer whose name I can't bring myself to write) receiving accusation after accusation of improper conduct by actresses he had come into contact with over the years and pressured into sexual acts. If all goes well, he will be gracing the inside of a jail before long, though knowing the broken justice systems around the world it will probably be some minimum-security resort with steak on Sundays and an adjoining golf course.

I recently spent an entertaining afternoon (okay, several afternoons) playing a video game called Red Dead Redemption 2 made by accomplished developer, Rockstar Games. A lurid and violent Wild West open world adventure game set in 1899, it nonetheless offered an interesting opportunity to immerse oneself in the history of the time, witnessing it, albeit through the lens of fantastical gameplay. One mission in the game involves escorting a band of suffragettes to a rally and culminating with the very satisfying pummelling of some local imbeciles who were particularly against the ladies in question obtaining any kind of rights.

This part of the game also provoked some furore when a dickhead youtuber who goes by the coward-handle 'Shirrako' took great delight in punching a suffragette and feeding her to an alligator. While the in-game mechanics could have handled this better, or at least regulated in some way the stupid actions of those who seem unable to know better, it is an open world game for a reason. The player is able to see that actions have consequences, even if only that people have to live with themselves. I wouldn't be surprised to find that Shirrako also posts on the Urban Dictionary...

While this action took place in a virtual world of make believe, a cornerstone of patriarchal society is violence against women. A United Nations study I read from 2012 summarised that in the outside world, murders are most often committed by men and against men, whilst indoors men are still the perpetrator, but women are most commonly the victim. I was once told by a policeman friend many years ago that in the UK, 90 percent of the murderers behind bars were men who had killed their wives. It's a frightening statistic, and as I'm sure you know, merely the tip of the domestic violence iceberg.

The sublimely talented Hayley Atwell in late 2018 recorded a video for Women's Aid that if you haven't seen it, I recommend you put this book/tablet/scroll down for a moment and watch it this instant; it's only a few minutes long and you will be all the better for it. It's a powerful, clever piece showing the emotional impact of physical abuse, and it stays with you a long while after watching it. I do give money to Women's Aid, which helps to support the victims of domestic abuse, but I think I'd secretly prefer if some of the money was used to put hits out on the ignorant sadistic shits who think it's okay to raise a hand to a women, or anyone else for that matter. Isaac Asimov once famously wrote via one of his characters: "Violence is the last refuge of the incompetent", and truer words are very rarely written.

Domestic abuse goes on throughout our society on a daily basis, particularly, but not exclusively in the lower income brackets. The frustrations of a life wasted away, the angry impulses that come with a general lack of self-control, the ache of living with decisions made by your younger self – a person you no longer recognise or identify with. So many reasons that apparently cause a person, typically but not always male to raise their fists against someone who trusts them, typically, but not always female. All wrong, in so, so many ways.

In early 2019, the shaving razor brand Gillette, in a desperate attempt to power themselves back to a position of prominence (or at least relevance) on the lips/faces of shaving enthusiasts released a new advert. It could be considered controversial, but only if you are an idiot.

It took the novel position that toxic masculinity is a bad thing, and that all, ALL us males should be doing something about it, whether we perceive ourselves to be at fault or not. We should police ourselves to start with, and also be looking out for, and stopping less clever (or more drunk) male friends from saying or doing things to women that might seem funny on a night out, but actually, in the cold light of day are at best rude and demeaning, and at worst make women feel threatened or insecure.

It was a brave move for a company hit by decreasing sales which I can only attribute to the rise of the hipster and their bloody neckbeards. They certainly wouldn't have made the advert out of the goodness of their hearts of course, but even if they are trying to hock razors, it's still a good thing to do even if it's a by-product. Some of the best things in the universe happen by accident after all.

They make an excellent point which, while obvious still somehow eludes prevalence in male culture. We SHOULD be calling out abusive behaviour wherever we find it. Gillette have long traded on the brand phrase 'The best a man can get', so switching up to 'The best a man can be' is hardly a figurative or intellectual stretch.

The result? Baying hordes of men, falling over themselves in misspelled tweet after misspelled tweet to vent their utter outrage at what they perceive as an attack on their masculinity.

Were I not already the winner of the most cynical person in the world competition for the third consecutive year (and eyeing up a fourth), I'd be putting my head in my hands. The 110 second long advert includes the phrase 'Some already are' to account for those of us (apparently a depressing minority) who already know that grabbing a woman's arse is a gross violation of their body and personal space that warrants prison time (ironically where you may end up on the receiving end of such gropes). Yet apparently there was still room for epic amounts of insecure men to feel threatened and take to social media to display their pathetic displeasure and inadequacies to the world at large.

At the time of writing, the advert has 608K thumbs up on YouTube, and 1.1M thumbs down. That's quite an outpouring of idiocy there, fellas. Keep it up, perhaps you'll stumble under a bus while being outraged on your smart(-er than its owner) phone and do a favour for the rest of us. If that sentence makes you put my book down in disgust, then fear not, it clearly wasn't meant for you. I have to hand it to Gillette's marketing department though, it's as if the Managing Director said, "I'd like to take a census of how many of our customers are idiots. Can you do that for me?", and marketing said: "No problems boss, leave it with us".

All told, things may be slowly, grudgingly, painfully dragging themselves in the right direction, but nowhere near as fast as it ought to. The reason is that all the old, mostly white males in key positions of power don't, deep down where it's rarely admitted to, especially on record or under oath, want things to change. Oh, they'll say all the right things, spout all the right platitudes, if they ever read my words would react with shock and outrage, but why on Earth would they want to? We live in a civilisation stratified around callousness, the meaner and more selfish you are, the higher you will rise. And once you have risen high enough, you get to make the decisions on behalf of everyone else. And what kind of decisions will they be? Callous ones of course.

The patriarchal society we've evolved ourselves into has so much to answer for, I scarcely know where to start. The #MeToo movement is an overwhelming force for good, though even then there is the occasional false positive. I was fortunate enough to go and see Aziz Ansari perform stand-up comedy early in 2019, and right towards the end of his segment, he addressed the allegations made against him last year. I should note at this point the allegations were heavily discredited when a great many female writers rallied behind him, pointing out that the main thing he was guilty of was not being a mind reader, and that the accusations themselves put back the #MeToo movement rather than progressed it. All Aziz would say was that he felt awful about how he made the lady in question feel, and that it made him question every encounter with a girl he had ever had, or likely would ever have in the future. A good guy trying to make himself a better man.

#MeToo is merely a recent footnote though in the history of patriarchal abuse. The Catholic Church has been repeatedly rocked by awful scandals of priests abusing children. According to many sources, some inside the church itself, as many as 5 percent of catholic priests have had a sexual encounter with a minor in the past 50 years. Oh, that's alright then, I thought for a minute there was something endemically wrong, but only 5 percent? Why didn't you say so, why that's barely 20,715 clergymen... Sex abuse in the Catholic Church is nothing new of course, accounts go back to the 11th century, where powerful men have abused their positions, and with it abused people they really, really should have been keeping their hands off of.

Moving to a more recent time though, patriarchy continues to thrive. In the modern business world for example, the gender wage gap is still alive and well. In 2018 alone, female journalists at the BBC still wind up being regularly paid less than their male counterparts, often despite working much harder in an attempt to earn the same amount. Not so long ago, women were still seen as property, and for some of those older males in power, I daresay they wouldn't object if our civilisation reverted to that model. It would at least give them all something to fight about while the world ends.

A sad fact that litters our species history is the viewpoint that women are seen as objects, to be owned and controlled, and not much has changed in the modern day and age when you consider the age and attractiveness of the average (significantly older) business mogul compared to his typical concubine.

I still wince every time anyone refers to the 'rule of thumb', supposedly named after a 17th century English legal ruling that it was legal to beat ones' wife, so long as the stick was no wider than your thumb. It turns out to be a complete and utter fallacy incidentally, no such ruling ever took place however it's a popular urban belief which no matter if true or not, should lead to not using the phrase, yet it is still regularly uttered.

Almost everywhere you look, the men rule, and the women do not. Other idioms like 'you throw like a girl' still litter our culture. Whether we hear it consciously or subconsciously, the message is clear: the female is weak, the male is strong. Most of men's favourite swear words (including my own I'm afraid to admit) are based on female genitalia – how messed up is that?

This is the world we find ourselves in, we have whether by accident or not (and obviously it's on purpose), we have crafted for ourselves a society where the strong rule the weak, the men rule the women. It's incredibly unfair, in fact it may be the most unfair thing there is, especially given the dark times that may await us.

But if one thing is clear above all else, it is that fairness is not a naturally occurring element on this planet.

Women didn't make the decisions that harmed our ecosystem to the point where we won't be around in the couple of centuries, but in a typically male fashion, we have made the decisions for them and are taking them with us on this final journey.

# CHAPTER 10 – UNSTABLE WEATHER

Unstable weather is something of a fallacy, and taking into account my innate contrariness, is a perfectly fitting way to start this chapter. The reason is that everything that happens within our atmosphere is inherently unstable. There are about a billion different variables going on, a computer powerful enough to model them all is a very long way away from being a reality. In fact, it's so far from being a reality that bearing in mind the long-term prognosis for the spaces, let's not start pinning our hopes on it arriving at all. We may just about get to flying cars though.

In the UK where I hail from, the greatest example of unstable weather in living memory, at least if you are of a certain age comes from 1987 and conjures up images of a saintly old weather forecaster for the BBC by the name of Michael Fish. On that fateful night of 15th October 1987, he stood in front of his blue screen wearing a trademark striped kipper tie and declared that earlier reports of a colossal storm bearing down on the UK were in fact a false alarm.

Needless to say, everyone went to sleep that night safe in the knowledge that we'd dodged that meteorological bullet. In the morning of 16th October, with thousands of trees felled, 22 dead and the National Grid in partial shut-down, the true story became clear. An extratropical cyclone (which strangely in the vernacular means 'not tropical' rather than 'additionally tropical') that would later become known as the 'Great Storm of 1987' had torn the UK and Northern

France a new one. Weather forecasting on the BBC would never be the same again. I personally completely slept through it, as I try to do for any and all tricky situations I may find myself in.

Every year we hear more and more stories of extreme weather occurrences around the world. Of Hurricane This and Tropical Storm That inflicting unbelievable amount of damage onto the infrastructure of our society, not even counting the severe death tolls they often bring with them.

In the early 2000s, some scientists started working on a new field of climate science they called Extreme Event Attribution. They had rightly got together and said to themselves, surely there is some kind of reason why the planet appears to be experiencing an accelerated amount of extreme weather phenomenon, and maybe, just maybe it's related to the amount of torment we humans are inflicting on our delicate ecosystem.

Since then, more than 230 extreme weather events have been closely analysed, from Hurricane Katrina to Typhoon Haiyan, from heatwaves and droughts to flooding, severe rainfall and everything in between. The goal of Extreme Weather Attribution is at once simple and immensely complicated; To determine whether there is a causal link between humanity driven climate change, and the extreme weather events our planet is experiencing. There are plenty of naysayers, as there always are; Pretty much everyone with a vested interest in unfettered capitalism is against Extreme Weather Attribution, and the concept of climate change in general. The kind of people with something that they don't want to lose, normally power and wealth. They will argue (or more accurately pay people to argue on their behalf) that the causal link is too difficult to prove, that the weather was always this bad, that the rising population densities around the world mean that the same old weather has a more damaging impact now than it did before.

For the past two decades researchers in this area have been developing tools, mathematical models and mechanisms of measurement in order to attribute beyond reasonable doubt whether extreme weather events worldwide are the result of climate change.

Their findings are unarguable, yet I could no doubt find some poor deluded souls prepared to do just that if I looked, and I wouldn't have to look very hard either. A friend of mine recently was told by someone he was winning an argument with on climate change 'I trust my instincts not the experts'. I believe this insanity was proffered by the Lunatic in the Whitehouse, whose name I am trying very hard to not use; I was unable to track it back to a particular quote, but it certainly sounds deeply plausible. Trusting instincts not experts must be a fascinating way to live one's life - I'm sure industries like Nuclear Power and Aviation would get along just famously if the decisions governing them were made on instinct rather than informed opinion. I'm certain there'd be hardly any meltdowns to speak of, or planes falling out of the sky minus their wings.

The Extreme Weather Attribution brigade chalk more than 150 of the 230 disastrous weather events down to climate change. To put it another way, these catastrophes that cause devastating loss of life, that have scoured cities and rural farmland alike, are indirectly manmade, something we have brought upon ourselves. They wouldn't have happened if we hadn't been so intent on filling the atmosphere with climate changing, greenhouse inducing gases, especially $CO_2$ for the past two centuries.

The tools developed by the Extreme Weather Attribution specialists to look for the smoking gun of climate change jammed into the back of our environment are now being rolled out globally. Regional environmental agencies around the world are now deploying their own Rapid Attribution centres, building on the work of these smart cookies, and with this, future Extreme Weather can be traced back to a single originating cause, us. While that is of course great, what is the point of being able to prove that climate change is involved? Surely, it's obvious to anyone with a couple of brain cells to rub together that even anecdotally (which is the level I operate best at), the weather our natural environment produces is getting worse, doing more damage and killing more people than ever before. What the heck do we think is the root cause? Sunspots? I should be careful making claims like that even facetiously, you never know when some bright spark will use my words to declare war on Sunspots.

When I was at university, many (an uncomfortably large number in fact) moons ago, I studied Astronomy. I didn't excel at it, the theories and practical telescopic astronomy have long been (and continue to be) utterly fascinating to me, but spherical polar geometry and calculating Hohmann transfer orbits will never be something I'm proud to say I'm good at, just like table football.

While I was there, I had an interesting argument with a lecturer regarding whether galaxies have black holes at the centre. He was a fine man, keenly intelligent and a personal friend of Sir Patrick Moore, who I now discover in my research was a right wing nutjob who said the BBC had been ruined by women. You should never spend too much time looking closely at your heroes, except of course if they were born with the name David Robert Jones.

In any case, the topic of the argument was this: at that time (late-1995) the understood wisdom was that it was thought that some galaxies 'might' have black holes at the centre. My arrogant contention was that it was blindingly obvious that all galaxies have black holes at the centre. My lecturer contended that however likely it might seem, it wasn't proven and given my grades and tenuous grip on astrophysical mathematics I was unlikely to be the person to prove it. I had to concede the latter points of course, but not the first – I knew I was right, even if I couldn't give him a cogent argument why.

It wasn't until some years later that I was proven right, but the point is not that I'm a smug git (though of course, I am), but the point is that some things you just intrinsically know given a few basic facts, and don't need to wait for the absolute proof, because by then it might just be too late to do anything about it. You find faith in the strangest places...

As the science of Extreme Weather Attribution matures and is rolled out globally, the theory is that governmental policy makers and will have a useful tool in their arsenal to help them with decisions that may adversely impact the ecosystem. The cynic inside me tells me that will be one more thing the policy makers have to ignore or simply state they don't believe in them, while being well funded by the corporations who back their political campaigns.

So then, while the root cause of the $CO_2$ emissions (spoiler alert, it has two arms, two legs and watches a ridiculous amount of reality TV) is comfortably covered in other chapters, would we benefit from better predictions of the weather? Certainly, back in 1987 if old Michael Fish had been advised to warned us to batten down the hatches a small number of lives would have been saved. All the trees would still have blown down though, for a supposedly powerful species we are utterly impotent and inconsequential when faced with the full fury of nature. Typhoon Haiyan that I mentioned earlier struck the Philippines in late 2013, doing billions of dollars' worth of damage and claimed more than 6,000 lives, which somewhat puts our Great Storm of 1987 to some shame. When I was much younger, I recall a study that found that weather forecasts for 'tomorrow' tended to be 40 percent accurate, whereas if you simply said to yourself that the weather tomorrow would be the same as today, you'd be right 60 percent of the time. Things have come on a long way since then, however.

The issue with long-term predictions of the weather goes back to the billion or so variables I mentioned earlier; anything more than a few weeks out tends to be wildly inaccurate. A fascinating new avenue of Weather Prediction is emerging to combat this, called Ensemble Forecasting. It takes advantage of recent advances in computing and numerical modelling to feed combinations of the key variables into multiple models at the same time, so instead of producing one often inaccurate prediction, multiple futures are examined in parallel. These multiple forecasts can then be compared and contrasted, where a trend appears across multiple predictions, it can be considered more likely to happen than something which is isolated to just one possible future. As computers continue to evolve and improve, so will the number of, and accuracy of the predicted futures, promising up to a years' notice on global scale events.

Moving out to the long-long-term, human-originated damage to the ecosystem is expected to see a rise of four to six degrees in average Global surface temperature. That might not sound like much, and full disclosure is the worst-case prediction, yet I somehow suspect the true cost will be much, much worse. In order to prevent an increase of this magnitude we'd have to completely curb our carbon and greenhouse gas excesses. Good luck on that with planetary politics focussed as ever on keeping a tiny minority in the manner they have become accustomed to.

In order to understand why a temperature rise of a handful of degrees is so bad, we have to travel down to the Antarctic, historically home to numerous scientific research bases, Nazi flying saucers and 'Things' from another world. Down at the ice desert at the bottom of our planet, research is being done on the historical state of our ecosystem. This is made possible through ice cores, drilling samples taken from ice sheets and glaciers in the region. The interesting thing about ice, is that at the time of its formation, some free air is trapped within its structure. And the ice of Antarctica is among the oldest in the world.

My favourite fact about Antarctica in case you are curious, is that at the Amundsen-Scott South Pole Research Station, on the first day of winter, when the last plane has left, and you are stuck there for seven months, the 45-odd scientists that make up the 'winter over' crew get together and watch the movie "The Thing". To make matters worse, it's a triple bill, starting with John Carpenter's 1982 magnum opus, moving onto the 1952 original 'The Thing from another world', and finishing with the 2011 prequel/remake. They are tempting fate if you ask me, though I one day hope to be there for movie day. To make matters even more surreal/spooky, a more recent tradition has sprung up whereby on midwinter, when they are halfway through their gruelling seven-month isolation, they watch 'The Shining'.

But enough about the crazy people who occupy Antarctica at such inhospitable times. By drilling for ice here where the sheets are more than three kilometres thick, we can tell much about the makeup of Earth's ecosystem going back over 800,000 years. One important discovery well worth noting is that there is a marked correlation between the amount of $CO_2$ in the atmosphere, and the planetary temperature at the time. Now that might seem obvious to some but try being a politician and you'll look anywhere but the obvious for a conclusion to draw. It's particularly important when you consider the billions of tons of $CO_2$ our species is so fond of liberating from the ground where its locked up and sending it into the stratosphere where is it free as a bird, albeit a good deal less tuneful. Secondly, and equally importantly, in the 800,000+ years of recorded planetary temperature this ice coring gives us, it's that is that the average planetary temperature of Earth hasn't been six degrees hotter than it is now. I'll grant you that going back a lot further it would indeed have been hotter when half the planet was still molten, but those conditions weren't conducive to life, were they. The point is that what is happening to our planet is manmade, almost certainly catastrophic and given our species-wide apathy, pretty much unstoppable now.

Adding just a few degrees to the temperature is expected to melt the remaining permafrost (frozen soil) around the world, which will release a metric shit-ton of greenhouse gas into the atmosphere, which will increase the temperature further, and trigger yet more feedback loops. As the ice shelfs themselves finally give up and melt, hundreds of millions of humans who live by the coasts will be forced inland by the rising sea levels. The increased temperatures will also make food production at any kind of scale next to impossible. Fresh water will become scarce. Country-smashing cyclones will become more and more commonplace. In short, dogs and cats living together times.

Much of the natural weather systems on our planet are caused ultimately by the planets rotation. Gravity has funny effects on free floating objects like, say, the molecules of air that make up our atmosphere. The name of these 'funny' effects (technical term coming up) is Coriolis Force, and on a rotating planet such as the one we sit so comfortably on right now, it has a curious tendency to push objects to the right in the northern hemisphere, and to the left in the southern hemisphere. I did try to write out an explanation for Coriolis Force for your delectation, but after two attempts I found I still couldn't understand it, suffice to say it has something to do with centripetal acceleration and nothing to do with the way water goes around in the sink.

In any case, this swirling combination of atmospheric movement gives rise to a certain amount of, well, swirling, which brings us on to our weather systems. When combined with solar radiation at high altitude, the Coriolis effect produces Jet Streams, vast, meandering air currents that convey heat around the globe, producing cyclonic (to read: big scary weather) effects.

As the climate changes, as the temperature keeps creeping up, we'll start to experience the effects of this unpredictable weather worse and worse. In 2018, an unseasonably hot summer set temperature records across Europe. In Japan, more than 20,000 people were admitted to hospital with heatstroke as the country experienced its hottest ever summer, with highs in excess of 41 degrees.

Meanwhile in California, you've probably gotten used to the wildfires that seem to rage every summer now, a careless spark or match giving rise to millions of acres of scorched earth. The wildfires of 2018 were the worst on record, far exceeding previous fires from way back in 1889. It was so bad that the California Department of Forestry and Fire Protection (Cal Fire) spent almost half a billion dollars on firefighting operations. Insurance claims relating to the fire, which consumed large parts of Malibu (and strangely $2.1M in cars on a car lot) topped $12 billion. I'd like to give a special shout-out to all-round good-guys Verizon, who during the wildfires took the courageous step of throttling the bandwidth of the Santa Clara County Fire Department data service once they had consumed their allotted 25Gb of data. Pay your bills and we'll let you fight fires was the loud and clear message from big business.

While the triggering event in each individual wildfire (some 8,257 of them in 2018) has a more immediate human causal factor than climate change, typically idiocy related to campfires, fireworks, smoking and good old fashioned arson, a massive contributory factor that would otherwise see thousands of these fires sputter out, is changing climate. There are an estimated 129 million dead trees standing in California, with millions more dying off each year. The causes are varied, but drought and extreme heat conditions are two of the most prominent reasons. Once dead, and without root systems to feed them nutrients and moisture, they dry out in the summer heat and once the worst occurs, they provide the fuel needed to turn small fires into incandescent blazes you can see from space.

For many years, the worlds climatologists did the decent thing when asked whether the increased extreme weather events were due to climate change and said diplomatic things like "It's impossible to draw a conclusion to attribute a single weather event". By 2018 though, it was clear that even not saying anything at all could be used against them. Friederike Otto, deputy director of the University of Oxford's Environmental Change Institute grimly highlighted: "If scientists don't answer, someone else gives an answer and it's usually people who have their own agenda,". Needless to say, Otto is now front and centre of the Extreme Weather Attribution movement.

The key factor in attributing Extreme Weather to climate change is that it's not the event itself that you measure climate change by, it's the likelihood of it happening. When I was growing up in the early 1980s, people still talked about the summer of '78 being a particularly hot one. Granted, I grew up in rural England and there wasn't much to talk about except the weather, but still, one got a sense that we didn't live in a subtropical paradise where such events were commonplace. I can even remember it snowing at Christmas one year when I was very small. The chances of a white Christmas in the UK now are infinitesimal, since due to the increased hot months, December is now better classified as early Autumn. I used to wonder why nobody talked about this, especially in a country that is so obsessed with talking about the weather, but we clearly don't talk about this because we don't want to face the truth.

Even now, with climate change on the lips of everyone who doesn't have a vested interest in the status quo it's still far too easy to turn over and watch something mundane on another channel. Out of sight, out of mind, the capacity to ignore what is right in front of us is one of our great skills as a species. It recalls to me the tale of Captain Cook's famous exploration of Australia in 1770, when the aboriginal natives completely ignored these hundred-foot long floating wooden monstrosities that had suddenly appeared off the coast. It's further explained in numerous contemporary interpretations that the ships were so alien to the natives' interpretation that they couldn't see them, or at least their subconscious chose not to see them. However, as anyone born with a disability in the modern world sadly has all too much experience of, being ignored doesn't mean you really are invisible.

During the 2018 heat wave, Friederike Otto and her team ran hundreds of simulations across two distinct models, one where we have added an awful lot of carbon into the atmosphere, and one where we hadn't, the pristine model if you will. When compared to recent Northern European heat waves that killed an awful lot of pensioners and had everyone else mopping their brow and saying banal things like 'how about this heat', the truth became clear. Climate change means catastrophic heat waves, which increase the risk of wildfires and kill trees to become easy fuel for the same fires, causing them to occur approximately twice as often. This is with a one degree rise in average temperature. When we're up to four degrees, or six, things are really going to start hitting the fan.

Surely human beings though, with all their big talk and big machines can take care of this? The idea of weather modification is not new, serious attempts with mixed results have been going on for more than 250 years now. Even earlier than that, a completely untrue belief that shooting prevents hail caused many towns dependent on agriculture to purchase and regularly fire cannons upwards with no shot in them. They'd have had as much luck (and spent a lot less) with pointing an admonishing finger at the sky, but in medieval Europe, even the Age of Enlightenment was a relative term.

Veterans of numerous wars during the 17th and 18th century reported rain that dutifully fell from the sky after every battle. I remember reading (but can't find a current source so it may have been in 2000A.D. for all I know) that during the 1980s, soviet artillery was used to disperse clouds before major military parades in Moscow, guaranteeing a blue sky. More recently (which I can verify) Russia used a technique known as 'cloud seeding', where chemicals are dispersed into cloud formations (typically delivered by plane) to force rain to fall earlier than it otherwise would, allowing you to keep a particular place at a particular time dry. The Chinese also deployed this technique to great effect for the 2008 Beijing Olympic games opening ceremony.

Going back even further in search of a weather modification origin story if you will, in 1590 the North Berwick Witch trials took place in Scotland. They centred around a vapid accusation that witches were behind some inclement weather that almost sank a ship carrying King

James VI of Scotland, while travelling to Denmark. It all kicked off when the admiral of the Danish fleet blamed the storm on the wife of an official in Copenhagen that he had apparently insulted – thereby doubling down on his abuse.

It really was a simpler time for your average cruel imbecile who loved ruining the names of good women. The accused while under torture named some more women, who named some more women. That might have been the end of it, but for old James, upon hearing about this witches coven uncovered in Denmark, decided that he wanted a coven of his very own in Scotland. Fanatical sanctioned idiots running around with pitchforks had soon snatched up more than a hundred innocent people, who all somehow, while being tortured admitted to devil worship and conspiracy to raise a storm to sink the kings ship; only then would the torture stop of course.

King James VI everyone, the first king of FOMO.

The militaries of the world have been having their fun with the weather control too, Operation Popeye during the Vietnam-US war is believed to have prolonged the monsoon season to disrupt the deployment of the Vietnamese army. Other efforts by many different military researchers centre around attempting to get an ability to create hail on demand. Nothing quite distracts a defending army after all like golf ball sized chunks of ice hitting them on the head.

While these are all well and good, at the very least we could consider them wholly ineffective considering the plight our species faces. What we might need, on a scale of planetary cooperation heretofore unachieved is carbon sinks. A carbon sink is a device or mechanism that sucks carbon out of the air and stores it. You know what the best natural carbon sink is? A forest. Too bad, we are busy cutting the last of those down, and besides, we'd need to plant literally billions upon billions of trees, on land we do not have to spare in order to start making a dent in the carbon problem, which in any case would take decades to have any effect. Another related problem is that the billions of tons of carbon we need to snag is up there in the stratosphere preventing reflected light from escaping (just like a greenhouse doesn't), some 10 kilometres above our heads, somewhat out of reach of your average tree.

Nonetheless, attempts go on, from the brightest of those who remain on the planet. The Bill and Melinda Gates Foundation is behind one such venture, another research team in Australia is perfecting an electrolysis technique that takes reconstitutes solid carbon from $CO_2$. Clever stuff no doubt, but, you know, stratosphere. 10 kilometres. So near and yet not really near at all, geographically speaking, or economically for that matter, all the solutions proposed work well in a lab with abundant power supply, not in a real world where every mechanism that exists to get anything of size into the stratosphere produces an awful lot of carbon as a by-product.

If there is one comfort of our torturing the ecosystem to breaking point, a great many scientists generally agree that our actions over the past few hundred years will delay the commencement of the next ice age by at least a hundred thousand years.

I seriously doubt our deep ancestors will be around to thank us for our foresight, however.

# CHAPTER 11 - HYDROCARBON ADDICTION

What's that, I pretend I hear you ask? You're not addicted to hydrocarbons – alcohol possibly, moaning about the weather very likely, Great British Bake-Off almost certainly, but certainly not Hydrocarbons.

What even are they? Something to do with fuel, right? Well the clue is in the name, they are compounds that contain hydrogen and carbon, two of the more common elements known to exist in the universe, and as such are among the simplest compounds out there. They are not to be confused with carbohydrates, which I admit writing by mistake more than once. The principle difference, and it might not sound much but it does make them very different, is that while both obviously contain carbon and hydrogen, carbohydrates also contain oxygen. Might not sound like much, but in biochemistry, it is of course a wide, albeit microscopic world of difference.

Hydrocarbons exist throughout nature; they tend to represent combinations of hydrogen and carbon that have been through an organic phase. Stuff that was alive once, however once it shuffled off the mortal coil, decayed and decomposed into hydrocarbons. In the early days of our planets experimental foray into life, this was an oceanic process. Early plankton analogues would die, sink to the seabed and be buried in sediment. Later, when life spread to the surface of Earth, the formation of hydrocarbons followed it. Everywhere, on land and sea, creatures were born, lived, fought, thrived and died in their untold trillions. The hydrogen and carbon that made up the formation of Earth, existing throughout the ecosphere, with more fed from below via volcanoes and hydrothermal vents began to be locked up via an organic stage into this new form. The humble hydrocarbon, awaiting a species to come along and add fire to the mix, without the awareness of what releasing all that carbon would do to what turns out to be a very, very delicately balanced ecosphere.

Human Beings have been burning hydrocarbons for fuel for thousands of years, in fact there is evidence that our forerunners, Homo Erectus, first started using it a little over a million years ago. While the actual origin is lost in the mists of time (having never been recorded in the first place), I imagine it went something like this.

Two Homo Erectus were walking in the woods one day. Don't all good stories start that way? They were young, barely adults, but at the current rates of life expectancy they were more than halfway through their lives, not that they understood such abstract concepts. They had language of a sort, but they were limited to the basics, food, threats and sex. Suffice to say soliloquies by Shakespeare it was not.

On that particular day it was cold, raining and they were thoroughly miserable. All the good game had gone to ground in the foul weather, pickings were sparse. Not more grass for dinner, one of them thought bleakly. The rain worsened, they considered turning back and returning to their tribe empty-handed but decided to continue onwards just a little longer, see what might be awaiting them over the next rise.

By the time they reached the next valley the weather had become stormy, thunder and lightning flashed and sounded respectively. Both men, if that's the right word to describe them, were startled by this new development of intense light and sound, but being seasoned hunters, it wasn't their first rodeo (if they had known what a rodeo was anyway) – they had seen and heard it before. They of course had absolutely no idea what it was, or where it fitted into their world view, only that it existed, and didn't seem to try and kill them outright, unlike many of the other creatures they often encountered.

Then, something unexpected happened. Lightning struck a tree in their glade, both men railed back, terrified and hastily reconsidering their worldview that the storm wasn't trying to kill them. The intense energy imparted in the lightning bolt, approximately one billion volts worth, caused the tree to catch alight, something neither of them had ever witnessed before.

Let there be fire.

One of the men turned and ran in terror, but the other man was uncharacteristically curious. Ignoring his impulse to follow his friend to safety he cautiously stepped forward. Then back again. Warmth was emanating from this strange new phenomenon; his skin felt the same way it did on a hot day when the blinding yellow light in the sky was overhead. It felt good. It felt comforting. Stepping closer again, he had to retreat, there was definitely an upper limit on how good being warm felt before it grew painfully uncomfortable.

The only issue was replicating it on demand. This is where the story gets a little hazy, or more hazy depending on your opinion of my conjectures thus far. Either this particular Homo Erectus had already observed the similarity of the tree fire to the sparks emanating from certain types of flinty stones when struck against one another and linked A to B to C, or the connection may have been made years later. Either way, the colossal power of the lightning and the secret it unlocked was soon replicated on a scale usable by our ancestors. It wouldn't have been without its missteps, definitely some fatalities, but they struggled, persevered, and tamed the flame.

What a colossal advantage that first tribe must have had over the others, still struggling, confined to caves at night during the winter. Fire was used for warmth and protection from predators initially, and later incorporated into food preparation. Little did they suspect that in just under a million years' time, the power they had harnessed would be used to wage war on the very systems and structures that provided them with life in the first place.

The Ancient Greeks, ever fond of a good yarn had a different origin of course, wildly untrue as all good stories often are, yet interesting enough to be worth mentioning. So the legend goes, fire was something only Zeus, Hera, Poseidon and the other gods could enjoy. They utilised it in their workings, especially Athena (Goddess of Wisdom) and Hephaestus (God of Fire), but to human beings this was strictly forbidden knowledge.

Enter the Titans, a second generation of deities who followed on from the classical gods but were a good deal less abstract, in real terms representing the greater knowledge mankind was obtaining about the world around him/her. Among the Titans was Prometheus, who became known for his fondness for humanity, and desire to elevate and progress our poor dumb old selves. This was of course some considerable time before he lent his name to a questionable addition to the Aliens movie franchise.

To that end, Prometheus stole fire from the workshop of Athena and Hephaestus and presented it to man, who were naturally thrilled with this new plaything, and put it to work right away with designing better means of killing one another. The punishment he received was severe, even by Greek God standards. Zeus sentenced him to be bound to a rock, where every day an eagle would turn up to feed on his liver. Because of his immortal status, the liver would grow back overnight, meaning the torment was unending. Harsh but fair I think you'll find, and I could with knowing how he did the trick with the liver.

In any case, however our forerunners obtained everyone's favourite exothermic reaction, the logical extreme was of course the Industrial Revolution, the intermediate stage between a simple pastoral life, and a more complicated existence where paying gas bills were now a thing. By this time mankind's tribal groupings had grown and evolved beyond all recognition, numbering in some cases into the millions or even billions. In order to support further economic growth (because if you've learned one thing about human beings by now, nothing is ever enough), fire was employed in all-new and in some cases obscene ways.

Which brings us up to somewhere near the present. As a species (and as individual countries) we're burning up hydrocarbons in the form of coal, oil and gas as fast as we can get our grubby hands on them. Over the past 200 years, since (you've guessed it) the Industrial

Revolution our expanding and developing species has got through colossal amounts of the stuff.

How much is colossal exactly? It's easy to doomsay after all, so here are some facts for you to set it in its right context. According to reliable scientific studies, in the last 100 years (as records are somewhat sketchy before that), our species has extracted and burned:

1.7 trillion barrels of oil

365 billion tons of coal

145 trillion cubic metres of natural gas

I hope those numbers scare you, because quite frankly they frighten the crap out of me.

Everywhere you look on planet Earth today bears the scars of this atrocity against the mothership, this megaproject to burn as much of the stored hydrocarbons as we can to fuel our 'progress', with scant regard for what it may do for our future survival prospects.

Surface mining of oil shale is a big contributor here. Thousands of open pit mines have been operated worldwide in the past few hundred years. Each is intensely damaging to the local flora and fauna. After a pit mine has exhausted its yield of oil shale, the mining industry half-heartedly reclaims the land, a great many suburbs of major Western cities now sit on reclaimed land, but the damage is already done, the landscape altered forever from the verdant pasture it was for so long.

Coal mining is even worse. The common practice of strip mining is devastating to local wildlife, and the land after is never quite the same again. The vegetation is gone, the soil destroyed. Any finds of scientific significance e.g. prior inhabitants, dinosaur, homo erectus or otherwise are generally decimated without question, indeed without even noticing most of the time.

As the near-surface deposits are running dry, new and innovative (by which I mean creatively destructive) techniques are deployed to get at those delicious fossil fuels, which bring with them new ways to mess with our delicate little blue-green stone. I am speaking of course about Fracking.

Fracking, or Hydraulic Fracturing to give it its full name, is the process of drilling a hole down into Earth, then sending some water under extremely high pressure down the hole. This pushes natural gas deposits out of the rock, back up the hole where it can be captured, and burned up to keep our turntables and Blu-ray players spinning.

You've probably heard that it's controversial in some way, protesters seem to find themselves frequently arrested while attempting to prevent the gas extraction, often by attaching themselves to heavy machinery. But what's so bad about it? Well, it turns out, and this may surprise you, but injecting water at high pressure into rock after drilling a hole all the way through it is an incredibly invasive technique that causes earthquakes. Please note, I didn't say it can cause earthquakes, I said it does cause earthquakes.

In 2011, drilling was halted at a test fracking site near Blackpool, UK after earthquakes of a magnitude 1.5 and 2.2 were recorded in the area. While earthquakes on this scale aren't exactly worthy of Hollywood disaster movie status, little is known about the cumulative effect of hundreds of small earthquakes on existing fault systems. Being as they tend to originate deep underground, we know precious little about Earthquakes, certainly not enough to be able to predict them with any accuracy, so deliberately causing them to me appears to be ignorance squared.

Oklahoma, a landlocked and previously geologically stable area with no fault lines, now records more earthquakes than California. To be fair, only a small proportion of these are due to fracking, but the rest have no more noble an origin story. The culprit is wastewater disposal, where fluid waste from oil and gas production is injected deep underground, deep enough they say to not affect our ground water, though I wouldn't for one be too sure of that. It all amounts to the same thing in any case, injecting water in the ground in order to, or because of our species' hydrocarbon addiction. An earthquake there in 2011 was measured at 5.7 on the Richter Scale, which is houseshakingly bad.

Energy generation is only half the picture though. A great deal of our material civilisation owes its heritage to plastics. Reading this on a laptop? Plastic in no small part. A kindle? Plastic. Paperback? There, I can't fault you, though the printer that it was printed on? Bingo.

Over the past 200 years or so we've been getting more and more into plastics in a big way to supply many of the solid objects that proliferate our consumerist culture. Sitting in my apartment writing this paragraph, little in my eyeline doesn't have some plastic components of some kind. Only the dining table and a few chairs can make any claim to the opposite, and I dare not look too closely at the screw fittings to discover that some plastic found its way into their makeup.

Two of the primary ingredients to artificial plastic, ethylene and propylene can only come from hydrocarbon deposits. I say artificial to differentiate the naturally occurring plastics in our environment, for example amber (fossilised tree resin) and latex, which comes from the rubber tree. Since they occur naturally, funnily enough they are in harmony with nature, they occur in limited amounts, and biodegrade naturally, albeit sometimes over extremely long timescales. A word of warning, in checking my homework, or at least finding out a little more about what I'm raving about, try not to google 'amber' and 'latex' together, you may be surprised by the results.

Let's continue on then with the unnatural plastics. Ethylene and propylene are first obtained by 'cracking' oil or gas, which is the chemical process of breaking down the complex organics of hydrocarbons into simpler forms which can be better manipulated. These are then further processed into monomers (small molecular forms such as styrene and ethene), and then polymerized (formed into polymers) to make final plastic products of various forms, shapes and sizes. All for the pleasure of your material consumption.

It is generally acknowledged that the first plastic created was Parkesine, imaginatively titled so by Victorian inventor Alexander Parkes. This material was patented in 1856 and won a Bronze medal at the 1862 World's Fair in London. This first commercial application of plastic amusingly enough was as an ivory replacement. You know what it's like, you love a home filled with ivory as much as the next person, but do we really have to shoot all those poor elephants and rhinoceroses to get it? Or perhaps it's just a bit too expensive. Look no further than Parkesine, for all your home synthetic ivory needs. I'm paraphrasing, but I'm sure the marketing campaign was probably a lot like that.

If plastic use had ended there, if it had gone out of style with ivory and a grand piano in every home, we'd be a lot better off now, but unfortunately our consumption of plastic has a great deal to do with our hydrocarbon addiction.

After World War 1, wartime improvements in chemistry were applied to commercial manufacturing, and a new wave of plastic products began to hit the shelves. Or began to be shelves I should say, among many, many other things. The early leaders of the plastics market in the 1920s and 1930s were polystyrene and polyvinyl chloride, better known as PVC. Polystyrene found widespread application as a cheap building insulator, and for packaging materials including endless coffee cups and the dreaded burger containers.

As a processed manmade material, polystyrene is extremely slow to biodegrade, long after human beings go the way of the Northern White Rhino, there will still be small polystyrene boxes blowing around on the toxic winds, adorned with suspiciously ornate red M shapes - though I should note in 2018 a certain fast food manufacturer finally committed to phasing them out. Despite this, polystyrene continues to be abundantly produced, with several million tons more of it produced every year, it is found increasingly throughout the environment, notably in the Great Pacific Garbage patch, as previously covered. And it's not alone, by one estimate I read, humanity have generated 8.3 billion tons of plastics from hydrocarbons, of which as much as 6.3 billion tons has now been discarded as waste, all building up cumulatively in our environment in landfills, microplastics and everywhere you find human beings in fact.

It is possible to recycle some types of plastic, but not all. In some cases, it's either not possible or costs far more to recycle than can be gained from the sale of the final product. It all comes down to money, which in actual fact is made from its own plastic polymer, so I find that highly fitting.

What are we going to do then with all these untold billions of liberated carbon atoms floating around our upper atmosphere? The fossil fuel industry oddly enough has an idea. They are punting a scientific technique known as carbon capture and sequestration (CCS) whereby the free carbon is once again collected and locked away within Earth.

The genie is not put back in the bottle so easily however, and by all the reports I could get my hands on, promoting CCS as a magic bullet for continued mining (and therefore shareholder profits) is at best inaccurate, and at worst criminal. Storage underground that would safely contain all the carbon we create for the foreseeable future has not been proven, nor has the hundreds of kilometres of pipes that would be needed to feed this fantastical reservoir. It didn't stop Norway, ever the progressive nation from trying though, as of 2018 they are capturing and storing a million tons of carbon every year. Sadly, this is a drop in the ocean, or a more appropriate metaphor might be a shovel-full of Earth into an open grave.

It is a sad and simple fact that when it comes to climate change brought on by carbon liberation from hydrocarbons, big business has no interest in dealing with it, but spend a lot of time and effort pretending that they do. It is commonly known that the most strenuous climate change denials always originate from lobbyists in the fossil fuels sector. This happens more often than not in the United States, supposed champion for truth and justice, though not so much these days naturally.

From 2015 to 2017, oil lobbyist and climate change denier Jim Inhofe chaired the United States Senate Committee on Environment and Public Works. It was in fact his second term, having previously polluted, sorry populated the position from 2003 to 2007. At one point he described climate change as 'the greatest hoax ever perpetrated against the American people'. In an esoteric and pathetic attempt to debunk climate change, he once brought a snowball with him into the senate chamber, fatuously uttering 'I ask the chair: do you know what this is? It's a snowball'. What this achieved I know not, though I am fairly certain that climate change is alive and well despite his insistence to the contrary.

It brings to mind a famous statement by Neil deGrasse Tyson when debating the validity of evolution over some kind of deity-based origin to our species: 'The good thing about science is that it's true whether or not you believe in it'. As always with NdT, he was right on the money, and the slightly smug smile as he uttered that devastatingly pure and simple epithet will always stay with me.

Jim Inhofe was succeeded in 2017 by John Barrasso, another bought and paid for climate change denier, who stated for the public record 'The climate is constantly changing. The role human activity plays is not known'. Bravo Barrasso, bravo. It's a damning indictment that the best thing I can say about his stupidity is that at least he didn't say it with a melting snowball in one hand.

On 1st of June 2017 the United States announced it would withdraw from the Paris Accords. The Paris Accords wasn't, I'm sure you are aware the right to French bread and garlic, but was in actual fact a global effort, signed at the time by 196 countries globally to keep the global temperature increase as a result of climate change under two degrees. The stated reason for the withdrawal was that it would undermine the US economy. No shit, Sherlock. It will undermine every economy in the world, since what we're talking here is an effective partial reversal of the damage done by and since the Industrial Revolution.

The idiocy doesn't stop there, naturally. In 2018, the US government held an event at the United Nations Climate Change Conference (COP24) to promote the use of fossil fuels in the fight against global warming. It would have been hilarious if it wasn't so pathetic. But you know me, I laughed anyway. The event promoted the use of the highly inaccurately named 'clean fossil fuels', suggesting that investing in these was an adequate safeguard for the future of our planet instead of, you know, not burning so many hydrocarbons. This was the second year in a row that the US government had attempted this charade.

Lobbyists in the US alone spend approximately 900 million dollars a year in bribing, excuse me, awareness raising (slip of the typewriter) various members of the Senate and House of Representatives who can further the goals of the fossil fuel industry, namely to prevent future legislation that limits their activities (to read: shareholder profits). Said lobbyists are even having some luck rolling back previous laws around emission levels and purification/detoxification of industrial by-products that similarly provide a barrier to shareholders getting new extensions on their mansions and that 17th home in the countryside they've always dreamed of.

This type of behaviour is sadly not atypical for the good old US of A. Whatever remains of the American Dream is a pale shadow of its former self. British Playwright Harold Pinter was long a vocal critic of US policies, in 2001 in his Nobel Price speech he issued this brutally honest rebuke: 'The United States has in fact, since the end of the Second World War pursued a brilliant, even witty strategy. It has exercised a sustained, systematic, remorseless and quite clinical manipulation of power worldwide, while masquerading as a force for universal good'.

In the United States and many other supposedly civilised places, the roots of these aggressive foreign policies lie with the Military Industrial Complex. I've always pictured the Military Industrial Complex as an abstract row upon row of planes, tanks and bombs – maybe it looks like that 'lots of guns' scene from the Matrix, that would be fitting, I think.

The MIC (because I'm sick of typing it rather than any affectionate acronyming) is an unofficial alliance between the military forces of a nation, and the civilian defence industries that supply it. You only have to look at the board of directors of any defence contractor in the world, and I guarantee you will find a number of senior retired members of the military.

President Dwight D. Eisenhower (and former five star general) recognised the danger of this alliance, whereby the motivations and even safety of a nation can be jeopardised by the desires and greed of the MIC. In 1961 in his farewell address, he stated 'we must guard against the acquisition of unwarranted influence, whether sought or unsought, by the military–industrial complex'.

Everyone immediately took him at his word and took steps to dismantle this potentially fatal union of conflicts of national interest. Nah, of course not. The MIC is alive and well today, and why do I bother to bring it up?

What does a war machine need to keep moving? Hydrocarbons? How many solar powered tanks are you aware of? How many wind-powered planes do you know of? Okay, to be fair there are gliders, but outside some limited stealth applications, the MIC aren't going to be able to drop a 15 thousand pound 'daisy cutter' (incidentally winner of 'most obscene weapons name' three years running) from a glider anytime soon.

There's no escaping the simple truth that the MIC, who control great swathes of government policies globally (and what they don't control they influence) are in love with fossil fuels and have no intention of moving away from them.

It's a tale as old as time. Well, as old as civilisation. Individual corruption and greed, monolithic corporations and quasi-government entities control the narrative and discredit anyone who disagrees. They will bury any dissenters with lawsuits, depositions, libel cases, and when that doesn't work, as has been proven time and time again over the centuries, they will actually bury the dissenters.

They want you to act like a good consumer and continue making the ruling oligarchy richer, everything is just fine the way it is, there's no need to change anything and no need to act. We'll be just fine, thank you very much.

With all that in mind, let's revisit my original question. Are you sure you're not addicted to hydrocarbons?

# CHAPTER 12 - NATURAL DISASTERS

In planning out the structure of this book, I found I had a lot I wanted to say about natural disasters, which while I'm sure would be entertaining (at least to me), seemed a little off the beaten track of our search for the Tipping Point. In fact, I was in the middle of talking myself out of the chapter when I started to consider the contributory factors to many of the disasters that have recently affected this planet, or indeed are coming up in the future.

This chapter then became an opportunity to make the case that we are slightly more culpable than you might at first think for the so-called acts of god that occur on this planet with disturbing regularity. So how exactly do we cause a natural disaster, or unnatural disaster would be more appropriate? Even with our arrogant and egocentric behaviour, we have to draw the line somewhere, don't we?

Let's start by looking at an actual, honest to goodness natural disaster, as a control subject, if you will. The operative word there is natural. As in nature. Nature has been around a lot longer than we have, and as often happens with the very old, she is very, very grumpy.

Earlier in my adult life I had a fleeting desire to become a Geologist. I even have a very basic degree in Geology, the kind you get from turning up to the minimum of lectures and being a contrary young adult who was very ungrateful to a family that supported him attending university. As you can probably discern from my various writings, I have more than a passing interest in the sciences in general. I find them all equally fascinating. I often wonder why I didn't become a scientist of some sort, I expect at the time I could have made such a decision, the lure of decent pay sat in front of computer screens was stronger than the prospect of spending three more years to get a PHD followed by mediocre pay sat in front of computer screens.

Anyway, it was somewhere during that time where among other things I was slowly and painfully learning how not to be a selfish little shit, I picked up a few nuggets on the subject of geology. One of which was that ultimately, I didn't want to be a geologist. I'm firmly of the opinion that the primary goal of university is to learn how to learn and prove to the world at large you can stick at something for three plus years without giving up and running home sobbing, something many employers find a valuable skill. As a by-product of that, you're normally sick to death of a particular discipline after studying it for an extended length of time and end up picking a career in a completely different area. I'm always mildly surprised when I meet people occasionally who work in an area directly related to what they studied.

It wasn't this 'familiarity breeds contempt' model though that put me off geology. It was what happened at Mount St. Helens, a volcano located in Skamania County, Washington State on May 18, 1980 at 8:32am and 17 seconds (Pacific Daylight Time).

Kerblam.

An earthquake, the culmination of two months of similar quakes caused the entire north face of the volcano to slide downwards, incidentally still the largest landslide that has ever been recorded. The slide exposed the high pressure and partly molten rocks underneath, releasing that pressure very abruptly. The lava exploded upwards, outwards, and northwards. A column of ash and debris was propelled 25 kilometres up, well above the boundary that separates the atmosphere from outer space. Mudslides formed from melted glaciers on the volcano spread in all directions, up to 80 kilometres away in some directions.

Fifty-seven people were killed directly and almost immediately by the eruption, including David A. Johnston a US Geological Service Volcanologist. His final broadcast 'Vancouver! Vancouver! This is it!' has a hauntingly professional quality that has always stayed with me. David was almost ten kilometres away from the blast, so hardly standing on the rim of the volcano dangling instruments over the side. Yet still the volcano took him.

Closer still to the explosion, photographer Robert Landsburg, upon realising his life would be forfeit by the approaching blast wave, continued taking photos until the last possible moment, then calmly wound the film back into its case, secured it in his camera bag, the camera bag into his backpack, then laid over the top of the backpack to protect it (successfully I might add) from worst of the oncoming superheated ash cloud. Despite these heroic tales of nobility and sacrifice, the dark side of human nature survived and thrived naturally. The very next morning scavengers were picking the area clean (including Johnston's belongings) to be able to sell souvenirs of the disaster.

Volcanology had always held a fascination for me, a quality that has stayed with me since my younger days has been shall we say a slight lack of risk aversion, and the best you could say about that was it keeps life interesting. That, however, was the end of my own personal fascination with geology.

Mount St. Helens might have been the deadliest and most costly volcanic eruption to occur in the history of the United States, but by those standards, wasn't anywhere near the worst volcanic eruption in the history of mankind, not even the most famous. Everyone remembers where they were when Mount Vesuvius erupted on October 24, A.D. 79, especially if you were unlucky enough to be living in Pompeii on the slopes, where pyroclastic flows immortalised many residents in their agonised death throes.

At the time of writing (early 2019) Mount St Helens is making some very angry grumbling sounds again, speculation is rife that we may be in for another eruption. Watch this space, but I recommend you do so from a very safe distance.

I'd also like to give a shout out here to the Washington State county of 'Skamania' too, though it turns out the county gets its name from the Chinook word sk'mániak, meaning 'swift waters', and not that the entire county were huge fans of Ska music. They may still be devotees of Madness and The Specials for all I know, but it would be purely a coincidence if so.

In an earlier chapter I've already talked about the ongoing slow-motion self-harm we're committing in the form of the Great Pacific Garbage Patch, and in my previous work I have talked at length about a few other not so natural disasters we've somehow managed to inflict on ourselves. I won't repeat them here, though I would like to draw your attention to a few that deserve a special place in our trophy cabinet of dumbest achievements, almost Darwinian in their exemplary lack of foresight. I present to you, resplendent in their idiocy, the fossil fuel related catastrophes of Centralia, Pennsylvania and Derweze, Turkmenistan.

In Centralia in 1962, a local landfill was set on fire by local firefighters (I know, don't ask...) in order to 'clean it up', as they were so instructed. An opening beneath the pit led to the abandoned coal mines beneath Centralia. They weren't abandoned because they were out of coal though, it just wasn't economically viable, there were easier pickings for the coal companies of the time. This fire set the coal deposits beneath Centralia alight, where they continue to burn to this day. 57 years and counting. These days, Centralia is a ghost town, populated by less than ten hardy (to read: insane) souls who have refused to abandon their homes to sinkholes and other geological instabilities brought on by the fire raging beneath their feet. Another fact that amuses me is that it was the inspiration for the setting of the video game series, 'Silent Hill', which depicts a hellish demon-infested townscape, but Centralia requires no supernatural influences to render it uninhabitable.

Onto Derweze then. The incident there is better known by its much more sinister moniker, the 'Door to Hell'. In 1971 soviet miners attempted to setup in this part of Turkmenistan to divest the locals of some natural resources, but the outgassing methane deposits made mining there unsafe. 'Fear not', their trusty geologists told them, 'We can have this burned off in no time'. I imagine at this point the click of a cigarette lighter.

The Door to Hell has been open, fires raging for 48 years now and still going strong. In fact, it's a major tourist attraction for the area now – which is something of a terribly faint silver lining.

Another major volcano in the USA is in Yellowstone National Park. To put it slightly more accurately, Yellowstone National Park IS the volcano. Also known as Yellowstone Caldera, it's the largest super-volcano in North America, actually an overlapping region of individual calderas covering an area just under 4000km$^2$. Now, you'd be forgiven for scanning the horizon of the beautiful, mostly flat lands of Yellowstone and saying 'Huh?'. The reason you can't actually see a volcano here is the same answer as your next question, which may well be 'what's a super-volcano when it's at home?'

A super-volcano is a large, flat structure that exists under the ground. It is a region where the magma rises up through the crust but is unable to break through to the surface. The pressure of all that magma builds and builds, until they erupt. Erupt doesn't do the event justice though. The Volcanic Explosivity Index (scientists just love to categorise everything!) defines a super-volcano as a Type 8, the biggest there is. The one-word description of a super-volcano eruption on the chart is 'Mega-colossal'.

Now, they don't happen every day of course – geologists estimate they occur every 50,000 years or so. This is also why you don't see evidence of them at the blast sites. The eruption leaves a huge crater of course, which is then filled in over the subsequent thousands of years. The last super-volcano eruption the world experienced was the Oruanui Eruption in New Zealand, about 25,360 years back. The crater that eruption left is now Lake Taupo, conveniently masking the geological devastation as tourists no doubt look around it and say to themselves 'Gee honey, this place is beautiful'.

Prior to that was the Toba eruption in Sumatra, Indonesia around 74,000 years ago, which according to the geological record was even more climactic. The amount of ash thrown up into the atmosphere triggered a global winter which lasted more than 10 years, and it's theorised that our ancestors at the time would have barely survived it, with plummeting global temperatures, shortages of plants and animals as they too suffered the effects. That really could have been it for the human race, over before it really got started. Barely a footnote.

Super-volcanoes occur above geological hot spots. A hot spot in geological terms is an area of the mantle (the layer of Earth directly under the crust on which you sit/stand/lay) where something weird happens, causing regular (in geological time) and massive eruptions to occur on the crust above. Please forgive my vagueness here, when I studied geology more than 20 years ago hot spots were little understood, and in the intervening time it seems not much has changed, save a few slightly more plausible theoretical models. Nobody has ventured down to the mantle outside of Hollywood movies, and within the movies they failed to follow proper scientific rigor and didn't even return with samples, which I think is a bit poor.

You can think of the planetary crust as the cooled off exterior of a hot liquid rock, floating above it. If you're interested to try a science experiment, go toast some marshmallows. Not too many though, I can speak from personal experience there is an upper limit before you begin to hate yourself. Hot spots are the mantle poking up through the crust at the same 'place' time and again. I use place in inverted commas, because the crust floats on, and slowly moves across the mantle, so the hot spot doesn't move, but over millions of years, our surface does. If you look at the Hawaiian Islands on a map, you can see a great example of a hot spot doing its work. You can even predict exactly where the next island is going to pop up at some point in the far future.

So hot spots cause super-volcanoes. We don't know what causes hot spots, but life is full of little disappointments so let's move on. Yellowstone, being a still active super-volcano is going to erupt again one day, scientists and doom-saying headline-grabbing newspaper articles think that one day is overdue by more than forty thousand years. The Yellowstone Super-volcano has erupted three times in its history, 2.1 million years ago, then again 800,000 years later (1.3 million years ago), then the most recent was 640,000 years later (660,000 years ago). Hopefully you see the anomaly there – it's now longer since the last eruption than the time between the previous two.

We'll likely have a little notice in the same way we did at Mount St. Helens, in that weird stuff was observed, geologists were on-site, and the local populace noticed unusual gas emissions and low-level earthquakes.

What will happen when it does erupt? All we can go on is what happened last time. Short term, it buried the areas that are now Wyoming, Montana, Idaho, and Colorado under three feet of volcanic ash. Longer term, the amount of ash thrown into the sky triggered an ice age which Earth took 10,000 years to come out of. At a very conservative estimate, hundreds of millions of people of course, would die. In the long run, should another ice age be triggered, the very survival of the human race would be in question.

On the bright side, I doubt reality TV would survive either.

We won't cause Yellowstone obviously, but I felt it deserved honourable mention in this chapter because of our overpopulation of the planet. As we have done everywhere else, we've filled the area around the super-volcano with unsuspecting people, ordinary decent individuals being born, living lives, getting married, producing and raising their young.

As a species we may not always be pulling the trigger, but we're extremely comfortable putting people in front of the gun, so to speak.

Our lack of foresight is not just limited to dangers from beneath our feet either.

I'd like to introduce you to Bennu. He's a real catch, slim, rugged and a bit of a loner. He may be getting on a bit but still very quick for his age. In fact, he's a 0.5-kilometre-wide asteroid in a similar orbit around Sun to us, so perhaps don't swipe right just yet, you may have less in common than you think. He's billions of years old, moving around the solar system at a heady 100,000 kilometres per hour. If that wasn't amazing enough, there is a one in 2700 chance that he'll be impacting Earth in the next century or so. While that might sound insignificant, consider that if he hits, it would be with the force of 1200 megatons. The Tsar Bomba, the most ludicrously large bomb ever constructed by humans exploded with a force of 50 megatons and was detonated way above Earth's surface to avoid any geological instability. An impactor naturally enough, would do what impactors do best, and the resulting carnage would be nothing short of apocalyptic.

A fair few of the disaster movies (and to be fair TV shows) to grace our silver screens pay tribute to the possibility that an asteroid may one day wipe the smile off our planets face, not that on the whole it's the planet that is feeling smug about anything. In these depictions, humans caught up in this life or death struggle to save the planet inevitably band together, setting aside old differences for the sake of the future of the species.

Just kidding, they bicker, infight, champion their own personal agendas and stab each other in the back the first chance they get. When the planet finally does get saved, one is left with the profound feeling that it happened more by luck than judgement, or at least because the writers had to make it so rather than for any credible feeling of worldwide cooperation.

When faced with creative efforts of this type I inevitably find myself rooting for the asteroid.

If you are ever interested in what effect a full-on planet killing, extinction level asteroid impact would have on our planet, I enormously recommend the CGI piece of absolute artwork by Anselmo la Manna created for the 2005 Discovery Channel Series 'Miracle Planet', which depicts in hauntingly convincing fashion the effect of such an impact upon our world. On YouTube you can find it set to the equally haunting track 'The Great Gig in the Sky' by Pink Floyd, which is inevitably how all great pieces of art should be viewed, try listening to The Wizard of Oz synchronised with Dark Side of the Moon sometime if you don't believe me.

In researching Tsar Bomba, I came across a crazy fact that caught my eye. When such a large, incredibly destructive bomb goes off, one thing everyone would agree on is you don't want to be anywhere near it. Not anywhere within about 800 kilometres in fact. This gave the Russians a problem for how to safely drop the bomb without killing the crew of the plane that drop it. The solution was found, use a colossal parachute, a $1.6km^2$ piece of fabric to slow the descent of the 27 metric ton monstrosity to the point where it allowed the crew time to escape. Not exactly – it was calculated that the crew would have a 50 percent chance of survival, and that apparently was acceptable for the Russian Air Force. A coin-flip to decide if the crew lived to never tell the tale (for reasons of state security) of what they had participated in. They did live by the way, but it was very, very close – the plane was apparently tossed around like it was made of paper in the ensuing shock wave.

Asteroid Bennu is not alone either, he's just one of many that careen ominously around the inner solar system, intersecting Earth's orbit uncomfortably close at times. The JPL Sentry System scans the current list of known asteroids for possibility of collision over the coming century. It currently keeps track of 75 objects with a rating on the Palermo Scale of >-0.6, which is a scale of risk for items that need very close supervision as they continue around their orbits. Without going into the complex detail of the scale, suffice to say it's a measure of how in trouble we are; the nearer the number is to 10, the less point there is paying off your student loan.

The highest Palermo rating ever given was for the catchily titled '(29075) 1950 DA', which in 2002 was assigned a value of 0.17 for a possible collision in the year 2880. Against a maximum score of 10 that might not sound like much, but trust me, on a logarithmic scale it's time to start rethinking your choice of home planet. By 2015, more accurate measurements of its orbit had been made, and the rating was downgraded to −1.42, which also rates a 'Phew' on the Tim scale.

The year 2880 might seem like too far away to worry about, and to be fair I agree, and besides, in the grand scheme of things, we shouldn't be thinking 'how can we stop it', but more "how can we warn the cockroaches, because we'll be gone?". It would only be the neighbourly thing to do, like warning the new occupants of your house about the person next door who likes to play Def Leppard late at night.

The 75 objects being actively tracked by the JPL Sentry System are just a drop in the cosmic ocean. NASA Planetary Defence Officer Lindley Johnson had this to say in 2018: 'The 50 million kilometres surrounding Earth include as many as 25,000 objects at least as long as a football field. About 950 of those are a kilometre in size or greater, big enough to cause a global disaster if they struck Earth'. Pause for a moment and reflect. No, not on the numbers, is Planetary Defence Officer the best job title you've ever heard in your life? I hope he has a really cool badge. Or a TARDIS. Or both.

While keeping an eye on the skies is clearly of paramount importance, the governments of the world cut funding for this area again and again. For example, in the last five years the Minor Planet Centre catalogued more than 17,000 near-Earth objects, all in eccentric orbits that could one day coincide with our own - almost half of these were of sufficient size to wipe a mid-sized city off the map. Due to budgetary cuts however, more than 900 of these asteroids have ceased to be tracked. They are still out there somewhere, and it would be a shame if we were to locate one of them again the hard way.

The Lunatic in the Whitehouse in 2018 actually pledged $100 million in budget to develop defences against this threat from outer space. I suspect though he may be more concerned about what countries he might also be inclined to point said 'defences' at should the mood take him. In any case, the money has yet to materialise, like hundreds of other election promises.

Should we discover one of these larger asteroids on a collision course with Earth, there remains little we can do to ensure our survival. If we had decades of notice I'd put our chances at 50/50. Years of notice, maybe 10 percent. Months of notice? Start panic buying beer before someone else beats you to the shelves. You never know though, there is bound to be some interesting stuff laying around in elliptical orbits that might be able to help, perhaps we could find ourselves dusting off some Reagan-era satellite-mounted electromagnetic rail-gun or chemical laser, quietly dropped off into orbit by a precursor to the United States very own X-37B military space shuttle that exists, but you've probably never heard of.

If it happens, so what? Earth can take it. Outside of the other planets locked into their separate orbits, there is nothing left in the solar system that could actually crack open Earth and reduce it to shards of hot rock, the worst that would happen is to restore the upper crust to its primeval molten state. Its current occupants however, would not be so lucky... It's happened before of course, civilisation ending events. A biggie of course was 65 million years ago, and it brought about the end of a variety of related reptile species that for more than 180 million years had enjoyed the run of the worlds green and fertile lands.

Enter Chicxulub, the big brother of Bennu. Conservative estimates put its size at 11 kilometres, possibly a lot larger. It impacted on 17th February, 65 million years ago, give or take a few hundred thousand years. Its impact savaged the climate, bringing about a nuclear-style winter that lasted for decades. During this time 75 percent of plant and animal species on Earth became extinct, including all ground-based dinosaurs. A few airborne species of dinosaurs survived, their descendants still fill our skies to this day, and pester children in parks for slices of bread. They were totally unprepared for Chicxulub, which makes me wonder who had cut the budget of the dinosaurs' spaceguard program?

I've already covered how the material that makes up Moon used to be part of Earth prior to a Mars-sized interloper sticking its oar in, but as no life existed at that time (that we will ever know of) it doesn't qualify as a disaster, because there was nobody around for it to happen to.

A more recent, and debatably manmade disaster is the Sidoarjo Mud Volcano, which took place in May 2006, and is still taking place. A Mud Volcano is a formation where boiling hot mud erupts, funnily enough. You probably didn't need me to tell you that. The reason it erupts is from seismic activity, most often from near subduction zones where one piece of crust is going underneath another one. Hot water from deep within the crust bubbles upwards, brought by methane gases, where it mixes along the way with mineral deposits, forming a solution which looks, feels (and I imagine tastes) just like mud. They can range from a few meters, right the way up to hundreds of meters high.

The mud volcano at Sidoarjo, Indonesia stands unique in the world in that it is the first mud volcano to have been triggered by human activity. It is widely believed that responsibility lies with the blowout of a natural gas well in the area drilled by Indonesian drilling firm 'PT Lapindo Brantas', although the company officials (and some scientists) still contend it was caused by a distant earthquake.

In the early years after the blowout, the volcano was spewing up a phenomenal 180,000 cupid meters of mud every day. For reference, that is about 15 times the volume of a Nimitz class aircraft carrier. Villages for kilometres around were buried in mud up to 40 metres thick, and a 700-meter-high volcanic cone formed. A desperate race against time to build embankments to prevent catastrophic overflow of hundreds more square kilometres of Indonesian country was ultimately successful.

Nowadays, it's down to around 20-30,000 cubic metres of mud per day, still a phenomenal amount, but it is definitely slowing down. The environment in that region of course has suffered tremendously, streams and rivers continue to be poisoned with the mud and toxic heavy metals from deep within Earth. While the company concerned was acquitted in 2009 of causing the mudflow, debate has raged ever since. Scientific experiments have since confirmed that the drilling is the most likely cause, as opposed to a distant earthquake that happened to happen on the same day.

Another situation that has unnatural disaster written all over it is the Guiyu e-waste dump site in Guangdong Province, China, which operated approximately from 2000 until 2013.

What is e-waste, some of you may wonder? It's the electronic waste our civilisation generates and no longer needs. The detritus of our technological advancement. In other words, last years' iPhone. All the manufactured electronic products that feed our consumerist desires end up in places like Guiyu when no longer required. Computers, TVs, laptops, tablets, mobile phones. When it's reached the end of its useful life, when it's been re-sold on to the last person in the economic pecking order, it ends up somewhere like Guiyu.

How does it get there? It doesn't load itself onto ships bound for the orient after all. Well, e-waste used to be thrown into landfills, until some smart cookie realised the heavy metals in the components were starting to get into the water table, and perhaps it would be a better idea to ship them off to far flung parts of the world where they aren't our problem anymore. Out of sight, out of mind, as ever.

The aforementioned smart cookie works for your average government, in case you were curious. That's right, many governments in the so-called developed world have been taking steps to preserve their local environments by paying foreign organisations to pollute theirs instead. This was of course illegal under the 2002 Basel Convention on the Control of Transboundary Movements of Hazardous Wastes and their Disposal. Everyone knows that. Unfortunately, this convention, like much of the legalities that would otherwise stop corporations from making a decent profit is hilariously easy to circumvent.

It's illegal to ship hazardous waste to China? Well that's no problem, we'll just slap a 'charitable donations' sticker on it and send it wherever we want. There is evidence most countries in the EU have engaged in this shady practice. Way to go team, what a great example to lead. High fives all round. Vote for something because it's right but do something different because it's cheaper.

At its height, Guiyu covered a 52 square kilometre area and employed 60,000 local workers. More than 100 truckloads of e-waste were delivered to the site every day. The workers toiled in harsh conditions for 16 hours a day, all to receive approximately $1.50 in recompense. As you can probably surmise, this area at this time was also almost uninhabitable. The ground wouldn't produce crops, the water was undrinkable. 80 percent of children in the area (many of whom worked at the site) suffered from lead poisoning, with pregnancy miscarriage rates well above average.

Another relevant example is Minamata Disease. A manmade disease, which is plumbing a new low in our quest to understand unnatural disasters. Discovered in 1956, it was first thought to be a new biological contagion not previously recorded, because of all the people suffering from it in a particular district of Minamata Bay hence the clever name for the disease. The symptoms were a neurological degradation, loss of motor control, difficulty speaking. Tests soon showed however it was caused by organic mercury poisoning.

It was ultimately found that the Chisso Corporation, who had operated a chemical factory in the bay since 1908 were dumping contaminated waste directly into the bay and had been continuously since 1951, more than five years at that point. To be fair, they'd actually been dumping waste into the bay a lot longer, but in 1951 a change in their chemical process meant they started using ferric sulphide instead of manganese dioxide. I know what you're thinking – a schoolboy/girl error, surely everyone knows that catalysing manganese dioxide produces organic mercury as a by-product? Apparently not, but you would expect the decision makers at a chemical factory to have some inkling, even in 1956.

By 1959 it was clear the chemical factory was to blame, despite Chisso withholding information on its practices from investigators, and generally doing everything they could to block the investigation, even going to the point of hiring shills to research alternative causes of the disease, all in the name of making a few more bucks while they killed the local populace and ecosystem.

Finally fessing up to what they had done, Chisso installed a purification system in 1959, even going to the trouble of throwing a party to celebrate its opening. The president himself drank a glass of water produced by the machine. It later transpired that the glass of water was not in fact from the purification system, which in actual fact did nothing to reduce the amount of mercury in the water and was obscenely described later as a 'Social Solution' to the problem. The factory continued pumping its poison into the water, despite everyone outside Chisso believing the issue had been resolved, and staggeringly bought themselves another eight years of factory production, and profit through the deliberate poisoning of more than 2,000 of their countrymen.

After being caught out again in 1968, a bitter litigation followed through the 1970s, where Chisso persistently attempted to avoid any blame for their actions. While it was no consolation to those that Chisso's actions had already killed, the survivors did receive substantial pay-outs, and the president of the company at the time, as well as the former plant director were both eventually found guilty of involuntary manslaughter for their criminal negligence and recklessness.

I could go on, but I think you get my point. I could talk about the Dams we have put in place across our waterways around the world flooding more and more land, but I won't. I could talk about the Banqiao Reservoir Dam in Henan Province, China that failed in a typhoon in 1975, and caused the deaths of 171,000 people. But I won't.

Looking back at our long and lurid history on this planet it's clear that a great many disasters, natural and otherwise have befallen mankind. Many of these, when you look closer at them have our fingerprints if not on the murder weapon, then at the very least on the shoulders of the victims, pushing them headlong into harm's way.

# Part 4 – Is There a Plan B?

This is one small step for a man, one giant leap for mankind.
Neil Armstrong

# CHAPTER 13 - TECHNOLOGY

Last time we were here, I singled out Automation and AI for a bit of a bashing, at how the reckless endeavours in this area were hastening our demise, but today I'm going to attempt to cast the net just a little wider.

Technology. The art of invention, feeding mankind's insatiable need to create. Those who paint create masterpieces, those who care enough use their creativity to heal the sick. There is another segment, who constantly ask themselves, is the way things are done now enough? These are the inventors. Could the way things are 'done' now just, in some nebulously intangible way, be made better? The word 'better' covers a wide range of needs, whether easier, simpler, quicker, less manual.

The word technology comes from the Greek word techne which appropriately means art and craft. It all started with the wheel around 5,500 years back, and if you'll forgive the easy dad joke, things went downhill from there. The pace was slow at first, a major blocker to innovation was the lack of widespread knowledge transfer coupled with the relatively short lifespan of human beings for much of their history. Skills and experience were passed down through family units and small communities, but between towns or villages just a hundred kilometres apart, the level of technology could vary considerably. With the invention of the printing press in the 15th century, it became much easier to share information, and much harder to lose it (where it would then await re-discovery as has happened many times).

Broadly speaking, there are two competing theories of invention. The 'Heroic' Theory, and 'Multiple Discovery'. The Heroic theory states that inventions and scientific discoveries are authored by unique, genius level individuals, the Da Vinci's, Tesla's and Galileo's of the world. The Multiple Discovery theory contends instead that inventions and discoveries occur simultaneously around the world by multiple inventors and scientists.

At first impression, Heroic seems obviously the answer. There must have been one person who had the idea first, perhaps they told a few people down the pub about it, but aside from royalties, they would remain the primary credited source, surely? Perhaps surprisingly, that proves not the case much of the time. Take Oxygen for example. This 18th century discovery is credited to Carl Wilhelm Scheele, Joseph Priestley and Antoine Lavoisier who knew of each-other, but independently determined the existence of Oxygen in Sweden, England and France at around the same time. Another great example is the crossbow. Over the past 2700 years, the crossbow has been separately invented in different parts of the world on at least five occasions now. Depending on the twists and turns of the next half a century or so, there may yet be room for a sixth.

The core of the multiple discovery argument is that inventions are a product of their times, that once a society has reached a certain level of advancement, they are more or less inevitable, making the actual discoverer somewhat less important than the general Zeitgeist that made it possible. This trend continues to this day, just look at the Nobel Prize, where for the sciences normally 2-3 unrelated individuals end up credited with a single discovery. The only difference now is that with the ease of communication our generation enjoys to distraction, it's almost impossible for the inventors in a field to have never heard of each other, no matter how independently they work.

It's important to remember that in a world so fundamentally

interconnected, all invention is standing on the shoulders of giants – achievement is not possible without the work of those who came before you. The bacterial research scientist didn't invent the microscope - even the microscope itself is credited to three individuals: Hans Lippershey, Hans and Zacharias Janssen.

With the commencement of the Industrial Revolution in the 18th century, things started to speed up, and haven't really stopped since. Such was the pace of invention in those early days, so many advancements were seen that Charles H Duell, Commissioner of the US patent office supposedly stated in 1899 'Everything that can be invented has been invented'. While the quote might be apocryphal, it nonetheless illustrates how far off the reservation we were now as a civilisation, the territory ahead of us dark, featureless and unknown with nothing before to compare it to.

I'm sure you're familiar with the phrase 'Necessity is the mother of invention', its origin goes back at least five hundred years, and is never more relevant and truer than in wartime. The numerous mechanised conflicts we've undertaken in the past couple of centuries, or at least since World War One have advanced our species to greater and greater heights of technical ingenuity.

For example, Fritz Haber and Carl Bosch, two enterprising German scientists at the start of World War One developed a technique to convert atmospheric nitrogen (which makes up 78 percent of our atmosphere) into ammonia, a compound of nitrogen and hydrogen, and an essential component for trinitrotoluene, better known by its short-form, TNT. Prior to this the German army had an extremely limited supply of ammonia for explosives, one of their main supplies was bat droppings from Chile, so they were not exactly swimming in it, so to speak.

Great, I'm sure you're thinking – it proved deadly for Germans opponents, but how did that advance the species? One of the other and arguably better uses for ammonia is as a key ingredient for fertiliser. Fritz and Carls technique, from its humble beginnings as an instrument of mass destruction that contributed significantly towards the deaths of the forty million people who lost their lives in World War

One, now feeds more than a third of the population on the planet.

Aside from its dark origins, ammonia must surely be considered a top ten invention for the good of the species, but as you probably come to expect by now, we have to look a little deeper than that. What the technique has actually enabled is rampant population growth to its now unsustainable levels.

The road to hell is paved with good inventions.

Let's you and I have a chat now about the humble smartphone.

I have a terrible memory. It's a fact. Later in life I expect I'll get diagnosed with something that explains it and all my friends say something like 'well that makes a lot of sense', but in the meantime, my smartphone operates as a functional extension of my mind. It tells me where I need to be and when, it remembers things for me so I don't have to. It tells me the route from A to B. It's filled with productivity apps of every kind, all in an attempt to live a more efficient life, which if I was honest with myself is more about slowing that inevitable decline as I get older and my body starts to fail me.

Perhaps I'm making all that up however, to justify my smartphone addiction. It wouldn't be the first time I've lied to myself, so I should at least entertain the possibility.

It turns out that it's all Simon's fault, those of you who know a number of Simons will know what I am talking about. In this case, the Simon Personal Communicator was the first Smartphone launched to the consumer market in 1994, though the term smartphone wouldn't find its way into the world's lexicon until the year after. It let you make and receive calls, send emails and faxes, contained a calendar, notes, world clocks, calculator. All the things we take for granted in any modern device. It was massive, you'd have looked like you were using one of those giant WW2-era US army walkie talkies, and if I'd seen you with one back then it would have earned a decent laugh I expect, but I was an immature teenager in 1994 so you could blame it on that perhaps.

Smartphones really took off of course in 2007 with the launch of the iPhone by apple. I still remember where I was the first time a friend of mine showed me it. The cleanness of the form, the utility of use. It felt like I was holding a piece of alien technology, it was so far removed from the awful practically analogue handset I now felt ashamed to take out of my pocket. So, what has happened in the 12 years since?

Distraction and addiction for one. I for one often carry mine around in my hand, to save crucial seconds taking it out of my pocket when the urge takes me to do something trivial, normally to write a note before I forget it, set a reminder to do something so I don't have to remember it, or merely to check my messages for the fourth time that minute. It barely sees the lock screen. Yet people still have the temerity to complain about the battery performance of these devices, while running them into the ground with task after task that simply cannot wait another second. If you are a driver (I am not), that distraction could even get yourself, or worse someone else killed, and despite laws around the world prohibiting said use, it continues to this day.

For many of us, at least a third of our day is spent in front of some kind of monitor screen for some kind of work. I'm staring at one right now as I write this, from the 54th floor of an apartment complex in downtown Sydney. They are ubiquitous and everywhere, you can't escape them. Assuming you all get the regulated amount of sleep, that leaves eight hours while you are awake to do other things. Take two more hours off to get to and from your place of business, that leaves you with six hours to enjoy other activities. If I look at my life, that would doubtless include watching TV, films, and staring at something on my beloved smartphone.

Let's face it, we spend most of our waking lives staring at some kind of screen, big or small. And it's harmful in all kinds of ways. Our eyesight for one. Do all of you take a break from looking at your screen for 15 minutes every two hours? I often don't get up until my bladder is screaming at me for attention, and I'm not the only one. I used to joke that on busy days back when I was a consultant, my hopeful dreams of getting a lunchbreak were downgraded to being able to go to the toilet.

Blue light as it turns out, is an absolute bitch if you experience it through the evening. Electronic light emitters of any kind, from light bulbs to TV screens emit blue light as part of their wavelength. Blue light tells our body that it's daytime, so floods your body with the right kind of chemicals to keep you awake and alert. If you've ever experienced that feeling when you're exhausted, but go to bed and are suddenly wide awake, you might look at the last TV program you were watching as the culprit. Many devices now, including Smartphones and PCs come with inbuilt blue light filters for those late-night activities.

Who among you has felt the anxiety of low battery, of knowing that any second now you will be cut off from communication with the outside world? I am often envious of those disorganised few who will shrug during an evening and tell you their phone has run out of battery when you ask them to YouTube the latest clip of a sneezing panda or a dancing dog. Or how about the phantom buzz, where you feel a notification vibration in your pocket, only to retrieve your phone to discover there was no notification at all, your brain imagined it. If what I'm describing here sounds like an addiction, there's a very good reason for that. And the people designing the devices and applications know that only too well.

Consider Facebook. Disregarding their blatant contempt for the idea of safeguarding our data, and prioritising profits over customer wellbeing, how do people get so hooked on having a social network that it's more important than having a social life? I did have to chuckle at the naïvely righteous indignation of friends and colleagues through the Cambridge Analytica scandal. I had to point out to them on more than one occasion, "if you use a website and you don't know what their product is, you're their product".

No, the real crime Facebook and their ilk commit is much more serious, and is very, very insidious indeed. It all has to do with a deep understanding of how our brain chemistry works, and how to exploit it to manipulate our motivations, the so-called 'dopamine hits'. Dopamine is a chemical in the brain associated with short term pleasure, interpreted as rewards for beneficial behaviours. In the context of our evolution, that includes eating, having sex, performing physical exercise.

What social media scientists (and make no mistake, behind closed doors this is absolutely a science) have tapped into is the realisation that while using social media platforms, rewarding you in the form of likes, shares and positive comments results in these dopamine hits. The pleasure you feel is short term, but you miss it when it's gone, which brings you coming back to the platform again and again and again. A recent study, to be fair it was commissioned by Fox News, so treat it with at least some scepticism revealed that Social Media dopamine hits are akin to the reaction on the brain when we snort cocaine. More than a little bit sensationalist, which is Fox News all over, but I can't help feeling there is a gram of truth in there somewhere.

It all comes in the timing of these dopamine hits, which the social networks have the power to do, because they control when you receive notifications. In 2017 another study proved that Instagram algorithms actually hold back your hits so you can receive them closely bunched together so that it's sizable enough to trigger a dopamine hit. Neurologically speaking, the trick is to vary the rewards schedule. If your brain gets used to when a reward comes e.g. every ten minutes like clockwork, it ceases to trigger the hit, so the notification algorithms have to take this into account. All very sophisticated, and deeply, deeply concerning.

Many doomsayers will propound that our twin obsessions with mobile devices and social networks are the end of society as we know it. They are certainly going to continue to be a big part of everyone's lives, at least while stocks last, and should keep us nicely distracted while the ecosystem collapses around us.

Gordon Ramsay, celebrity chef and master of the coarse insult proudly announced in 2018 he had recognised the disproportionate amount of time he was spending on his mobile phone, and now no longer possesses one. While I salute the sentiment, I feel compelled to point out that with the money to afford multiple personal assistants (who you can bet all own multiple smartphones), I question why it took him this long to do so.

A close personal friend of mine has recently opted for an interesting hybrid approach, he recognises the need for communication using tools such as WhatsApp, but has downgraded to a non-smartphone which retains the capability for such communication, with none of the app-driven fiasco around social networks and dopamine notifications. I'll be watching his experiment closely, it would well be a gap in the market for a phone smart enough to keep modern forms of communication alive, but dumb enough to protect us from our own natures.

Speaking of WhatsApp, in 2014 they took the unusually selfless move of introducing end-to-end encryption, which is significant considering it has more than a billion daily users. This was great, unless you were a government that liked spying on people. I'm not for a minute suggesting there aren't good reasons to keep an eye on some individuals. Terrorists will terror after all. I won't comment on the fact that if you look back far enough in a terrorist's radicalisation you normally find a western country that committed barbarous acts first for their own economic gain; no more than I just did anyway.

There have been occasional claims in the press that WhatsApp has a backdoor, a way to bypass the encryption keys and view the messages sent between users as plain text. I'm happy to go on record here and say of course it has a backdoor. Of course it does. The backdoor would have been added as a result of extremely quiet but equally extreme political pressure, very possibly secret laws passed in secret courts in the countries that WhatsApp or their parent company Facebook like to incorporate themselves in. I have no proof to offer except my deep, bone-weary cynicism, all I can say is that when some whistle-blower inevitably leaks evidence of the backdoor, I for one will be very, very unsurprised.

On the radar for me now at least is an app called Signal. It's an instant messenger app. Great, you think, I've already got about a million of those. Well, it's encrypted. Same again you say, as far as you can prove anyway. The kicker is, it's owned by a not for profit foundation so it can't be acquired. Keep talking, you mutter…

Signal owes its heritage to RedPhone and TextSecure, apps

launched by Whisper Systems back in 2010. While they are owned by Twitter, they are 100 percent autonomous, and show all the hallmarks of respecting true privacy. Every system can be compromised, given enough care and attention by security agencies like the NSA or GCHQ, but I'm going to put my money, or at least instant messaging here for the time being.

Technology has even greater impacts on our health though than tricking us into biochemical addictions too. Obesity for one. A life spent in front of a computer screen tends to be synonymous with unhealthy snacking and reduced amounts of exercise, and ultimately if you're not careful high blood pressure and a heart attack. Lack of human contact also brings about depression according to several studies, which makes sense given we are inherently social creatures. Staring at screens, big or small, all day every day will lead to degraded eyesight and a need for lenses earlier than you otherwise might have.

Speaking of poor eyesight, I'm often amused by the amount of innovation that the pornographic industry has enabled. Michael Bucchino, senior interactive producer at Droga5 said 'Porn arrives at ideas faster than other industries', and he is absolutely right. The ability to make online payments came about because of the need to sell access to pornography online. Video streaming? Same again – the code to stream videos over the internet was first developed by a Dutch porn company in 1994. Thanks guys. Instant messaging came about because of a need viewers had to communicate with early camgirls and request their favourite sex acts.

Right now, Virtual Reality is finally gaining a foothold in the world thanks to the efforts of this very same industry. For true online innovation and a glimpse at what is to come in the world of digital technology, you need only look at the latest features released on Pornhub.

Security is another area where each new innovation is instantly met with misuse, if not outright criminality. Got a webcam on your computer? I strongly suggest you cover it up with electrical tape right away – only a low-tech hack is going to beat those who would take over your webcam and record you. Who would want to take over webcams? The governments of the world, for one, hackers for two, and we know very well which of those two groups is the least trustworthy.

In 2013, Edward Snowden, then an IT contractor with an exceptionally high security clearance disclosed several thousand documents on a top-secret US Government program called PRISM. It turns out that in the years following the World Trade Centre terrorist attacks, a series of secret laws had been passed that compelled Organisations such as Google, Microsoft, Apple and Facebook to turn over live feeds of all their data on everyone to the US Government. There wasn't a damn thing any of them could do about it, no matter how twisted, wrong and amoral it was, these wild invasions of our privacy were now officially legal. That's the world we live in now, any illusion we have of privacy is exactly that. An illusion. Seen through a prism.

Anything you submit online, be it text, images, videos or sound is copied and uploaded for analysis by the governments of the world and their pet machine learning algorithms. All things being equal, their goal sounds reasonably noble on the surface – to prevent terrorists from doing what terrorists do. But all things are not equal, not by a very, very long way. Do you know and trust the analysts working for your government? Do your views align with your government? What about the other governments they secretly share data with? The recent furore with Chinese conglomerate Huawei concerns a very ugly rumour that their technology distributed around the world may be feeding confidential data back to the mothership, for furtherance of China's global agenda. I can't speak for that, but I can share a personal anecdote that you may find interesting.

In 2010 I did some work in Shanghai. I had been to Hong Kong a few times, but this was my first trip to the mainland as it were. My hopes of a traditional Chinese cultural experience were dashed when the only place I could find to get a coffee to alleviate my jet lag was of course a Starbucks.

Security in Communist China was naturally a concern. I wasn't up to anything particularly secret, especially in the fairly narrow segment of marketing I inhabited where we were just trying to do something slightly cleverer than the competition. A colleague of mine on the same trip advised me to use a VPN before attempting anything online, especially from the hotel internet. Why I scoffed, were the Peoples Republic of China interested in me, a lowly IT Consultant? Turns out they were. Not that I was anything special, they were simply interested in everyone.

On the last day my colleague was in Shanghai before leaving me there to complete the project, he gave me a pro-tip over a beer. Write down the serial number on the bottom of the Internet Router in your hotel room before you leave for work tomorrow, he informed me, smiling lopsidedly, enjoying the idea that he knew something I didn't. I sipped my beer and eyed him sceptically – was he on the level? After a few years of working together, him knowing something I didn't wasn't exactly an everyday occurrence. Perhaps he had started drinking a lot earlier than me? Nonetheless, my curiosity got the better of me and the next morning I duly took note of the router serial number. I went out to work and it was soon out of my mind. That evening I returned to the hotel, and sure enough, the serial number was different.

The make and model were identical, but the actual router had been switched out for another one during the day while I was out. There was no innocent explanation – those things run for years without need any kind of maintenance. The only valid conclusion was that it had been recording my actions on the internet (while it was encrypted through a VPN, I've no way of knowing whether even that was really secure) and the government had collected that data as a matter of policy. My digital footprint from that trip, encrypted or otherwise, likely remains stored to this day, another brick in the Great Firewall of China.

The days of digital privacy have come and gone. And that's just the official abuse of technology, illegal activity like identify theft is on the rise too. With the proliferation of data breaches that happen around the world on an almost weekly basis, sophisticated criminal enterprises that may or may not be state sanctioned (North Korea, looking at you) are stealing record amounts of money. In 2017 in the United States alone, $16.8 billion was stolen last year as a result of identity theft, a number up from the previous year, and while the figures aren't out yet, 2018 doesn't look to be any less lucrative for the hackers.

The stolen data doesn't stop at monetary gain either. I mentioned Cambridge Analytica earlier in this chapter, their clandestine efforts to influence the US elections and Brexit have yet to be definitely proven, but I'm hoping it's a matter of time before the figures behind these travesties of societal manipulation see the inside of a jail cell. Hope is not the same as optimism though, that's not the kind of world we live in. Russia are being forced to deny their involvement in manipulating the 2016 US elections on an almost weekly basis. At the time of writing in early 2019, the whole world is still holding its breath to see if there is any fire to back up what amounts to a phenomenal amount of smoke.

On a more individual level, cyberbullying is on the rise too. The twisted desire to hurt or embarrass others for ones' own amusement definitely makes the cut as one of my favourite reasons that human beings are my least favourite species. You don't find aardvarks or hippopotamuses going out of their way to treat their fellow creatures with such disdain. According to the US-based National Crime Prevention Council, 43 percent of teens have been victims of cyberbullying in the last twelve months and 81 percent of the teens surveyed believed that others cyberbully because they think it's funny. Funny. I've written before of the cruelty of children, so I'll spare you the repeat, though abysmal parenting of course deserves most of the credit for turning out the little shits who grow up thinking it's normal to prey upon others.

Another area on the rise are the Artificial Intelligences infiltrating our daily lives. While some fear the AIs will one day take our jobs, others fear an even more sinister uprising will take place one day. The jury remains out. Bill Gates believes that AI will be good for humanity, while Elon Musk believes they will become our overlords. I believe Elon, but then I'd be the first to bend the knee to our new silicon overlords. I don't find myself fearing a possible future where incorruptible, infallible and omniscient beings take over from the walking lie factories that hide behind the word politician in our current society. Things certainly couldn't be run any worse, an entity with no concept of money or desire for its acquisition would be a breath of fresh air in the disgusting world of modern politics.

That more or less brings us up to the present day. Why don't we take a look at what's going to happen next? The signs are there for me, that the big thing I see on the horizon is greater take-up of Virtual Reality, and the beginning of a retreat from reality. After all, what has reality ever done for us? The societal collapse envisioned in Ready Player One (2011) written by Ernest Cline is likely a close blueprint for what will happen over the next few decades, except perhaps the obsession with eighties pop culture – that feels a little bit too much like wish fulfilment.

We're already seeing the first signs of this today. In Japan, they even have a word for it: hikikomori. It is a word to describe reclusive adolescents who have shut themselves off from the world, their only contact with anyone being through the internet. It is estimated there are more than half a million of them, living solitary existences away from sunlight while their parents likely surf the same internet, looking for a way to reach them. Addictions to technology are another warning sign. Video Game Addition is a well-established mental disorder, as is addiction to the grey wafer of silicon you carry everywhere with you – I am very likely a card-carrying member of this group as you have no doubt already surmised.

Freely available evidence points to an energy crisis in the coming decades. The world of plenty as we know it will come to an end. It will limit our ability to travel as fuels become prohibitively expensive. Intercontinental travel will once again become something only the rich can enjoy. Cargo transportation will undergo the same impact. The food you eat will be grown within a few hundred kilometres of your location at most, drastically reducing the variety of your diet – I just hope for your sake you don't live in a sprout area.

Artificial intelligences will continue to supposedly support humanity, while in reality all they support will be the wealth of those who employ them. With so many people out of work many countries will be forced to adopt a 'basic income' approach, where everyone's basic needs (food/shelter/water/internet) are met regardless of whether you have a job or not. The governments of the world won't like it, but they would like a 60 percent homeless problem and a collapsed economy a lot less I'm certain.

The reduction in free movement between countries will enforce more isolationist viewpoints, further feed the regression inwards, as people stay indoors more, finding new and more innovative ways to kill time using the internet.

So, there you have it, dear reader. The lack of available energy in the coming decades coupled with increasing digital crime (both legal and illegal), and a return of the ugly isolationist tendencies that made the middle ages such a fun place to live make it all but certain that technological advancement will not be the panacea that saves us from ourselves.

# CHAPTER 14 - WHY ASTEROID MINING WON'T WORK

It's more or less at this point that I'd like to digress and talk about David Bowie. I like to do it in every book I write (two at the current count), but it feels like a nice tradition to establish. I don't think I've quite digressed with the first sentence of a chapter before though, so that also qualifies as a new record in discussional deviation, if I may be so bold as to propose it for Olympic recognition.

You may recall a dedication to David Bowie within the words of my previous book, if indeed you stuck with it that long. I talked about his effect on my life, largely unrealised until the moment he was no longer among us. Now, there's no easy way to put this, so I'm going to have to come out and say it, to confront the elephant standing on top of the elephant in the room as it were.

Since Bowie travelled beyond the rim on 21st January 2016, this planet has become an increasingly unpleasant place to live. There, I've said it. I don't say this in the 'a shining light went out of the world' kind of way, though it's true of course. No, I'm speaking in a much more literal sense. A series of calamities have befallen this planet, the rocky orb we falsely believe we own, and are terrible at sharing with others. I'm serious. The two major ones are evident to everyone who hasn't been living in a hermetically sealed vault somewhere underneath Wyoming for the past few years. I don't know anyone who did that by the way, I'm just being typically figurative and overly florid.

I of course am referring to Brexit happening in the UK, and the Lunatic in the Whitehouse. There was the usual parade of shootings in America, plane crashes, bombings around the world, but these two events will be the ones looked back on as cultural Tipping Points to accompany the ecological ones already well underway. Fear not, I shan't be caught talking politics except in the usual general sense to highlight the complete and utter untrustworthiness of those who seek public office. Suffice to say, bad things have happened, have kept on happening and will continue to happen, since Ziggy Stardust returned to Mars.

It got to the point where the popular news parody website NewsThump noticed. They published a memorable article six months on from the dark day with a truly amazing headline, poignant, funny, and uncannily true all at once: 'Reality continues to crumble in the wake of David Bowie's death'. It drew the same conclusions that I have come to, that the further we are from 21st January 2016 (1101 days at the time of writing this sentence), the worse off we are. Look up the article, it's well worth a read, as is much of the content on NewsThump.

I didn't know David Bowie personally, never met him, in all probability our paths would never have crossed even if he'd lived another thirty years, but I couldn't disagree with the articles central premise, that we as a species haven't been able to keep our shit together since that day forward. I'd be the first to agree that it's an unusual correlation, but you can't dispute the central fact that things for humanity have taken a distinct turn for the worse since the Thin White Duke departed us.

I recently watched a film. Point of fact, I recently watched many films. It's what I tend to do when I'm not working, thinking, writing, eating or sleeping. It was 'Valerian and the city of a thousand planets'. Based on the long-running epic French science fiction comic 'Valérian and Laureline', the movies opening sequence did something for me though that no film had ever done before or since. It opens with a montage of sorts, showing the organic growth and evolution of a space station bearing a remarkable similarity to our own International Space

Station.

A deeply symbolic series of handshakes take place over decades as Earth's various peoples one by one make the long climb to the stars, docking their diverse vehicles with the Space Station (soon renamed Alpha) and greeted in this way before adding to the station, both materially and culturally. Space Oddity plays over the top of the entire scene. Perfection, I thought.

The sequence took an even more fascinating turn as the arrivals continue, now a parade of increasingly bizarre aliens who arrive on Alpha to call it home. Each time they arrive, another smile, another handshake. The ultimate symbol of unity and friendship used again and again to welcome myriad species into an inclusive future. Tropes were averted beautifully, fierce, warlike aliens greeted the human ambassadors in the same way. Perfection squared.

As incredibly amazing as this opening sequence was (I reminded myself of it as I typed this passage), it made me unbearably, inexpressibly sad. That was all. A profound, soul-deep sadness that soon passed with the next film and the next, but I couldn't stop thinking about that emotion, it wasn't something I had ever quite felt before, and as always, the new fascinates me.

Turning my thoughts inward, I soon had the answer. It was because I know, fundamentally that this future won't happen for us. No matter how fantastical a film ends up being, if it starts on Earth in familiar surroundings, you can't help but feel a connection to it. No doubt Luc Besson, master craftsman that he is, knew this all along. That was the reason for my sudden attack of the blues. This utopian vision of our upcoming inclusive future among the stars won't happen, can't happen. That beautifully shot montage was the closest we will ever get to a future, that if we could have started thinking and acting as one, could have been within our grasp.

It wasn't sadness at all. It was grief. Mourning for a future that I knew could not be.

Apologies for the sombre note, but it felt relevant enough to warrant inclusion. Let's change tack, maybe even get back on point for the chapter, though I think I managed to pull a link out of my backside there that I'm quite pleased about.

Asteroid Mining then. Why don't we start with an obvious question? Why off (not on) Earth do we need to careen through the vacuum of space to dig up some hunk of rock way out in the middle of nowhere, Texas?

It all comes down to a simple concept, scarcity. One of the many facets of our untenable occupation of our pretty planet is that we use more resources than we can lay our hands on. In the grand, non-human-centric scheme of things, this is a problem that largely rights itself over geological timescales. Some people came along, they used all of X, whatever X was, it doesn't really matter, but they needed X in order to survive. Those people aren't here anymore. Even the X they used up will end up being renewed as rocks are eroded, crusts subsumed and deposits that were impossible to find before coming to light.

You know I'm talking about human beings obviously, you're not stupid. So, let's assume we actually want to solve the problem of scarcity, at least some elements of it, in a slightly less final way for the two-legged animals that stumble around staring at grey rectangles of silicon for much of the day.

Earth Overshoot Day exists as a symbolic measurement of at what point during a year does our civilisation consume more resources than can be naturally renewed. It looks at factors like carbon sequestration, growth rates of food, wood and other naturally occurring resources, offset against how much we need for 7.5 billion people and growing. At the moment, the Overshoot happens in early August, but is creeping earlier and earlier every day.

That means that each year, by August we have consumed more resources than are available. We are then on credit for the rest of the year, by which I mean we are borrowing resources from the next year, and the year after that – it doesn't take an economic genius to know that state of affairs can only go on for so long before there are no more years left to borrow from.

Based on current estimates, the human race requires 1.7 Earths to maintain its population and societal structure. Sadly, nobody is making any more Earths, which means we are slowly sinking deeper and deeper into a future we cannot extricate ourselves from.

Earth you see, is a closed loop. By that I mean no more resources are coming into it, so that closed loop is all we know, yet it doesn't stop us as a species from consuming resources in an orgiastic frenzy. All of which means the loop is only getting smaller. We could have curbed our excesses, prevented uncontrolled population growth, limited resource consumption of course, there were enough dissenting voices in the scientific community making the rest of humanity aware that this was happening. We could have, but we didn't. Coulda, woulda, shoulda...

Asteroid mining is an idea to avert that catastrophe. It turns out, following the formation of the solar system, a lot of free-floating chunks of rock are still out there. You may have heard of the Asteroid Belt between Mars and Jupiter, less well known are the Trojans, a concentration of asteroids in Jupiter's orbit, following it around Sun like a lost puppy.

Occasionally one of these lumps of rock hit something, which for the smaller ones we call a 'meteorite', and for the larger ones the term 'planet killer' is generally deployed. In between hitting things, they just float around the solar system doing absolutely nothing. Being basically leftover building blocks from the formation of the solar system, they are formed from the same elements that used to exist abundantly on Earth before humans started using them up.

It's a lovely long shopping list of stuff our civilisation loves to slurp down. Water is an obvious one, many of these rocks are partly, some predominantly water-ice. And who couldn't do with more fresh water? That's just the start though. Going down the list, depending on their makeup type, the asteroids can contain all kinds of goodies, such as gold, iridium, silver, osmium, palladium, platinum, rhenium, rhodium, ruthenium, tungsten, iron, cobalt, manganese, molybdenum, nickel, aluminium and titanium.

This makes the asteroids a very tempting target indeed for a nearby planetary culture being choked to death by its own excesses. And they are all just floating around out there, ripe for the picking, not owned by anyone or anything. The more egocentric amongst our current space pioneers consider them our legacy, man's inheritance if you will.

The idea of fetching resources off-world isn't new, however. In 1898 American astronomer and writer Garrett Serviss spoke of 'the yellow gleam of the precious metal appearing under star dust' in his book 'Edison's Conquest of Mars'. His fantasy scenario was of wealth acquisition rather than planetary salvation, and let us not kid ourselves, the current asteroid mining enthusiasts are not particularly altruistic. They would not be funding research into these areas without an expectation of a serious return on investment.

It remains an exciting prospect though, brave pioneers in their spacesuits landing on asteroids, looking around at a featureless, grey plain, and unhitching their pickaxes. They would probably be robotic in nature though, the idea of sending roughnecks to asteroids will remain the province of Hollywood only.

The two corporate front runners in this race to escape our closed loop are 'Planetary Resources', backed by Larry Page and Eric Schmidt of Google and James Cameron of every good science fiction movie, and Deep Space Industries, a more discreet endeavour without any of the glitzy charm.

Should they get off the ground, what would be their destination? They aren't short of targets, but the smart money is on Ryugu, a kilometre-wide Near-Earth Object, recently visited by the Japan Space Agencies very own Hayabusa2 spacecraft in 2018. I did idly wonder if this meant the Japanese could claim it as their territory, but apparently not. The International Outer Space Treaty (1967), prohibits any nation from claiming 'sovereign territory' in space, so the route for private enterprise is clear. After all, when has a country ever ignored treaties to get what it wants?

And what's so interesting about Ryugu? Quite a number of things as it happens – it's large for an asteroid, one kilometre across and weighing 450 million tons in fact, but not so large that gravity would make landing there a problem, or even fuel expensive.

It's in a stable orbit, it's not plunging around the solar system every 76 years like Halley's comet. It's near, in cosmic terms. Still hundreds of millions of kilometres, but that's a walk in the park when compared to a trip to, say Saturn.

Lastly, and definitely not leastly, the estimated economic value of Ryugu for mining is purported to be more than to be 80 billion dollars, having an interesting theoretical composition of, amongst other things nickel, iron, cobalt, water, nitrogen, hydrogen and ammonia. Everything a growing civilisation needs.

Going beyond Ryugu, ambitious plans talk about colonising Ceres, the largest asteroid in the asteroid belt (so large it's actually a dwarf planet in the same category as Pluto) becoming some kind of frontier town that acts as a hub for transporting mined resources back to Earth, or even Moon or Mars if needed. If that sounds like the basis for a science fiction TV show, there's probably a very good reason for that.

That all sounds great. Why won't it work then?

Strictly speaking, we're talking about two different questions here – why it won't work, and why it won't help us. We'll start with why it won't work.

There are a lot of hurdles to clear in order for asteroid mining to become a reality.

First off, as dull as it sounds, we have the legalities. I mentioned the International Outer Space Treaty. This was a decade long piece of bureaucracy finally signed off in 1967. It was kickstarted of course by Russia's launch of Sputnik in 1957. This was the true beginning of the Space Race, and many countries around the world were terrified that that race would end with everyone in space, possibly on Earth also, speaking Russian.

The treaty stated that outer space would belong to the 'province of mankind', and all nations would be free to use and explore said space, provided it was done in such a way as to benefit all mankind. If that sounds vague, it would have been intended so, with obscure, hard to pin down legally phrases like 'province of mankind' and 'benefit all mankind'. Every nation signing it would have their own opinions on which loopholes they might jump through in the years to come, but in just a short decade of wrangling it was the best everyone could agree to.

In theory, it might leave the door open for private enterprise to mine and own resources, BUT it doesn't explicitly say that, so it remains effectively untested regulation in this regard. Certainly, since the end of the US / USSR arms race, the subsequent collapse of the USSR, the destruction that befell not one, but two US Space Shuttles (Challenger and Columbia), government interest in space exploration, particularly in sending men and women up out of our gravity well had waned somewhat, so there is a vacuum to be filled, pun intended.

Since then we had The Moon Agreement in the early eighties. You may know it by its slightly catchier name, 'Agreement Governing the Activities of States on the Moon and Other Celestial Bodies'. This permitted the extraction of natural resources from off-world, but no space-faring nation would ratify it, because it also banned all kinds of fun activities like military use, weapons testing and claims of property. It also required all research conducted by any country in space to be publicly available to the entire world. I can easily picture Ronald Reagan and Mikhail Gorbachev crumpling up the treaty and throwing it across their respective offices in the White House and Dom Pravitelstva, respectively. Both missed the waste-paper bin, of course.

More recently we've had regional efforts, such as the US 'Space Act of 2015' which did make the owning of resources obtained off-world explicit, provided your organisation and company directors were based in the good old US of A (and thereby paying taxes there). Since then, The Grand Duchy of Luxembourg (I love its full name, I find it hilarious for some reason) has surprisingly weighed into the debate, passing their own law on the exploration and use of space resources in

2017. The Luxembourg approach hopes to attract asteroid mining concerns to base themselves there by being more flexible and allowing company directors to be based wherever they want.

From a legal standpoint, despite these more recent efforts, the legal path is still murky. Once again, we're hoisted by our own petard of being unable to co-operate at a global scale. Whoever first launches their mining equipment into space and takes aim at Ryugu, they will be doing so without a cast-iron certainty that they will definitely own the resources they are spending billions of dollars to accumulate.

Then, there's the technology required. Any miner will tell you there are many different ways of mining, each with their own advantages and challenges. When you factor in equipment that must work in a vacuum, low to zero gravity, function while clogged in dust you can't see through that just won't settle, you start to appreciate that this is going to be very challenging indeed.

Currently, it costs $22,000 to get a single kilogram from Earth's surface into orbit. It can technically be done significantly cheaper by using SpaceX reusable rockets, but the going rate is still the going rate. If you somehow strike a deal with Elon to use the rockets at cost price, you'd still be looking at upwards of $10,000 per kilo.

A top of the range mining excavator, say the Liebherr model because we want nothing but the best, weighs 116 metric tons, which even at a discounted going rate would cost $1.1Bn to get into orbit. I'm not suggesting for a minute we start wandering around a Liebherr showroom with our credit card looking for equipment to take to Ryugu, but I hope you start to see my point. Mining is rough, difficult, heavy work, whether you're in the Australian Outback, South Africa, or downtown Ryugu. Equipment that will last is going to weigh a fortune, so to speak.

A Space Elevator would save us a lot of time and money, able to ferry cargo up and down at a fraction the current cost, but no such megastructure exists. Materials science has not yet developed a mass producible material the elevator could be constructed from that could withstand the enormous stresses put on such a structure. Carbon nanotubes have long been mooted as the possible solution to this intractable problem, but a great deal more R&D needs to go into the construction processes before this can become a viable material.

In 2018 Japan actually launched a prototype Space Elevator into orbit, I was surprised to discover while researching this chapter. Before you start packing your best party outfit for a trip to Low Earth Orbit though, the test was a 6cm robot traversing a ten-metre cable between two satellites, but it did go some way to proving some theories about the effects of elevator travel in a vacuum, which are definitely questions I'd like answered before setting foot on a real one.

It's a long way off becoming a reality however, the most optimistic (i.e. unrealistic) estimates say that a Space Elevator will be in place by 2050 – I'd say 2080-2100 would be more likely, if not even later. To be absolutely fair, the companies behind these statements are Japanese, and nobody, nobody does megaprojects like the Japanese. What else would you expect from the county already busy converting great swathes of mainline train tracks to maglev.

I have to admit, while it does still have innumerable blockers against its feasibility, a working Space Elevator could be the gamechanger we need, not just for asteroid mining but the future of the species itself. If someone told me that humanity could survive to reach the stars, but we'd all be speaking Japanese, to that I'd have to say Konnichiwa.

Moving on with our great Trek to the Stars (see what I did there?). Say we get there to Ryugu, with all our equipment intact, and ready to go. Gravity, which was our friend on the descent to Ryugu, then becomes a bitch once operations begin. All forms of mining create a lot of disturbance and vibration. Our hypothetical mining machinery will require anchoring to the surface of the asteroid at all times to avoid being flung away by Newtonian Laws of Physics. I mentioned it earlier, but in the low gravity, dust clouds kicked up by machinery will linger for days, reducing visibility to near-zero. In this harsh unforgiving environment accidents will be commonplace. We're going to need a lot of redundancy in what we bring. More equipment, more weight, more cost.

How are we going to control the mining equipment? Ryugu is a few minutes away from Earth by fast horse radio signal. We could ship humans along, but we'd not only be significantly adding to the cost to bring an oxygen shell for them but also massively adding to the risk profile of the mission. Alternatively, the machinery needs to be autonomous. Perhaps it's time for Artificial Intelligence to stop playing chess grandmasters and Jeopardy! and start earning its keep. I'd be very sceptical of the success of this, because an AI, to be any use at all, needs to be trained. To train an AI to recognise an apple in an image, for example would require thousands of images of apples. For an apple. We're talking about deploying sophisticated and dangerous machinery to an unknown environment and expecting it to just know what to do. We'd frankly be incredibly lucky if a human being, trained for years if not decades for the mission managed to achieve anything without something going badly wrong.

Say the mining goes well, and we get a few thousand tons of goodies ready to be shipped back to Earth. It would need to be something on that magnitude to justify all the cost, and that's just to begin with. First though, we've got to get them off the surface of Ryugu. Not a problem, we bring with us, or construct from raw materials an electromagnetic rail gun. We can shoot packages back into Low Earth Orbit. Careful with the trajectories here though people, it's won't be like knocking on the neighbour's door to ask for your ball back should one of those babies accidentally impact Earth. A safer route might be to aim for lunar orbit and use that as a staging location.

As fiddly as I made that all sound though, that's the easy part. By successfully mining an asteroid and returning the refined materials to orbit, we're fine and dandy for the foreseeable future of constructing stuff in space, we can use the materials to build ships, space stations, hell even space habitats if we want to. How do we get it back down to the surface though, where if my dire predictions for the next hundred years come true, there is going to be the most need?

The Space Shuttle had a cargo capacity of 22 metric tons, so if they were still in service, it would take 45 missions per thousand tons of ore. That assumes the Space Shuttle could even handle re-entry carrying that much mass. Since Vostok 1 launched Yuri Gagarin into

Space in 1961, all Space Vehicles built since have an interesting factor in common: the pull of gravity leads to an inbuilt design bias towards launching large masses into space, and return small masses to Earth, typically just people in a small container. It's easy to see why of course, we're stuck down here at the bottom of the Gravity Well, mankind dreamed of escaping it for centuries if not thousands of years, and we're still only 58 years on from that first deathtrap that took Yuri where no man had gone before.

Heatshields & parachutes for those volumes of resources will be impractical. Aerobraking is another option to get things down to Earth, skim the atmosphere to shed velocity, allowing payloads to come in for a softer landing. I say softer, we'd still likely be talking a terminal velocity of 53 metres per second. Still, the payloads of raw materials could probably survive that and still be usable, so perhaps we designate some empty desert somewhere, paint a giant bullseye and retreat to a safe distance.

That's a long, convoluted and hopefully broadly informative if not entertaining answer to the posited question of why I don't believe Asteroid Mining will work – there are too many problems in too many areas. I've been wrong before, but now we come onto the second question – why Asteroid Mining won't help us, and you'll see that even if I'm wrong about why it's not possible, it probably doesn't matter.

What you have to understand however, is that we are at war. Our opponent is intractable, merciless and cannot be reasoned with. The enemy is for once not ourselves, at least not directly. It's time. All the problems I've painstakingly listed we could resolve, as individuals at least we make fearsome problem solvers. The inescapable conclusion, the conclusion of the five odd thousand words here, is that our civilisation is running out of time.

The time we do have left is not being used well. As we speak, more humans flood out of wombs all over the world to consume more resources, the atmosphere is being polluted, the oceans poisoned, the forests cut down, the globe being warmed.

As a megaproject, asteroid mining might conceivably fix our global economy if we desperately wanted it to, but not our broken ecology, our dying environment, an influx of new resources from the solar system would actually exacerbate those issues rather than go any way towards resolving them. That's something our old friend the Lunatic in the Whitehouse won't appreciate, since he's never let facts get in the way of good old-fashioned making money.

I saw a recent article from researchers from Yale and Harvard proposing a radical new way to tackle global warming and climate change. Dim the Sun. I did first write Dim Sun there, in line with my Sun naming convention policy, but I thought it might be confused with a certain delicious Chinese appetiser, and also it made me hungry. Feel free to step out for a snack if you need to, I know I did.

Ahem. Anyway, yes, you heard me right. My geeky brain instantly went to ways Sun itself could conceivably be forced to dim, put a giant mirror in between us and Sun, alter Moons orbit, or even something more esoteric and fantastical like dropping a wormhole into the corona to siphon off great swathes of the energy it generates.

Their proposal however is a bit more realistic and less science fiction, yet somehow no less absurd, which is a pity really. Their proposal is to use either naval guns or giant tanker aircraft to flood the upper atmosphere with sulphate particles which would block sunlight and therefore reduce the amount of heat getting through. Less heat getting through means less being trapped, which could in theory lessen the impact of global warming.

I honestly didn't know where to start on this boneheaded, ass backward approach to the problem at hand. It does the sum total of nothing to address the actual problem, that our civilisation is continuing to poison the planet with greenhouse gases that are causing this runaway effect. It's the scientific equivalent of sticking your fingers in your ears and humming a loud noise when you don't want to admit how stupid you just sounded. I can only hope some Professor at one of those venerable universities delivered some swift retribution in the form of failing grades for the researchers.

I would like to give an honourable mention though to the Chang'e

4 lander which early in 2019 proved it is possible to grow plants under controlled conditions on Moon. To be fair there is no reason why it wouldn't work since it was using Earth soil, and it's already been proved in zero gravity on board numerous space stations since the seventies, but it is nice to have a little more confirmation. I'm certain Elon Musk, with his Mars colony plan quite possibly humanities only credible lifeboat is taking detailed notes.

So that is Asteroid Mining. It has a vanishingly small chance of succeeding, and in the event that it did, does nothing to curb our own excesses, and what we are busy doing to our own ecosystem. The only thing asteroid mining might conceivably do is make it easier to bootstrap our collective asses out of the gravity well, what with so many resources already being available out there in space, untouched and unspoiled. Perhaps the carbon in an Asteroid like Ryugu could be spun into a handy nanofilament and lowered down to form the Space Elevator for our species to flee up, perhaps the iron could be smelted to form starships to carry us out among the stars.

And perhaps Moon is made of cheese.

# CHAPTER 15 - THE FUTURE OF TRIBALISM

Tribalism? As in what we used to be like, before we started wearing suits and staring at smartphones more than we look at our loved ones? Well, yes and no. Yes, as in yes, that's what we used to be like, and no, as in – there's no before about it - we're still very much in that mindset, whether you know it or not.

For almost the entire history of humanity we've been tribal creatures. For the past few million years, right up until the last tiny sliver of time, human beings have spent most of that time in small groups, rather than alone or in large groups. The maximum number on those tribal groupings tends to hover around 150 according to various social and anthropological studies, as well as a very interesting theory called Dunbar's Number. In 1992 Robin Dunbar posited the widely accepted sociological theory that there is a cognitive upper limit to the number of people we can maintain relationships with, or at least care a damn about, and he came up with the same number – 150.

More than 150, and people fall out, conflicts take place and groups reshape with smaller numbers. Less than that, and things tend to fall apart without enough people to fulfil all the various functions a group might need (foraging, nurturing, defence etc). Tribalism is the sweet spot in between these two outcomes, a stability our species has always subconsciously strived for. Which led me on to something rather interesting, a correlation I don't think anyone has drawn before, which means you'll be the first to read it.

Ever heard of the Lost Colony? It's an unsolved mystery from 1587 in an area of America which is now Dare County, North Carolina. You may know it by another, more ominous name if you know your history – Roanoke.

The colony at Roanoke didn't go well from the start. Sir Walter Raleigh had ordered its establishment in order to further English strategic interests in the region. To that end, he assigned his friend John White to become Governor of the colony, which would consist of 115 English settlers in search of a new life. While White had been to the island before on a previous expedition, he was primarily an artist, and not well suited to the job of administrative governance. This is why we don't let our artist friends hold anything more dangerous than a brush or a gin and tonic.

When the colonists arrived, they could find no trace of the English garrison left there the previous year, the discovery of a single skeleton did not indicate a happy ending. Spooked at the missing soldiers, the colonists initially declined the kind offer to colonise the island but the captain of the ships that had ferried them across the Atlantic would not let them back on board, so there they stayed.

Relations with the Native Americans were shaky, many suspected they'd had a hand in the disappearance of the garrison, so trust was difficult to come by. On both sides as it happened, as for the native inhabitants, the memory of Aquascogoc, a village burnt two years earlier in retribution for the alleged theft of a silver cup was still fresh in their minds. Yep that's right, a missing cup was all the evidence the colonials needed to burn a village to the ground. It's a sick joke that European colonists perceived themselves as a civilising influence on the indigenous populations, and points to a conclusion I have posited before – inside every famous explorer from antiquity, is a psychopath dying to get out.

Desperate and short of food, the colonists begged Governor White to return to England to ask for help, and he did so, leaving his wife and daughter behind, promising to return as soon as humanly possible with fresh supplies. However, just as he arrived in England the Anglo-

Spanish War broke out, and all ships were immediately commandeered to counter the threat of the Spanish Armada. It was over three years before White was able to return to Roanoke. He found the settlement deserted, no trace could be found of the 90 men, 17 women, and 11 children. There was no sign of a struggle, the people were just gone, apparently of their own free will.

From there things get really vague. The word CROATOAN was carved into a fencepost, the name of a nearby Island, but White was unable to investigate further because of an oncoming storm, and his men refusing to go any further, so the next day they departed. Really? His wife and daughter missing, and he does the most cursory of searches before returning to England empty handed? Subsequent expeditions did not fare any better, and relations with the Native Americans deteriorated to the point where the Europeans were attacked on sight, making further investigations impossible.

There are many competing theories as to what happened to the lost colonists, the Spanish attacked, they moved to parts unknown and subsequently starved to death without being able to reconnect with their English patrons, they joined a Native American tribe and moved away, the same Native American tribe massacred them. And those are the conventional theories. Among the more esoteric theories, they reverted to cannibalism, some other lost group had reverted to cannibalism and had them for dinner, they were the victims of witches, they were killed by the mythical Native American spirit monster, the Wendigo.

The point I want to make is that 115 is significantly under the Dunbar limit, but well into the region where the group risk being unable to function effectively. While any of the theories are possible, some are significantly more plausible than others. Left to their own devices for several years and likely losing more of their numbers, conflicts would have been commonplace, exacerbated by the shortages of food and other essential supplies. A larger group of people would have retained greater stability, but dysfunctions would have forced the colonists to fracture into smaller and smaller groups, until no social cohesion were left. It's likely under those circumstances that several of the (conventional) theories are true, some joined the Native Americans and migrated away, some starved to death while seeking a better site for a colony. I'm pretty sure a Wendigo wasn't involved though.

Back onto tribalism then. Over the last few centuries we've been moving into unknown territory as a species, migrating into larger and larger communities, some of which number in the low-double-digits millions (aka cities). This is of course, from a tribal perspective (and one could argue many other perspectives) highly dysfunctional. Yet some of our tribal roots remain. We are tribal in our way of thinking, and in the loyalty to our social group above all others. This leads me onto an inescapable fact about our modern tribal society.

We're loyal to our tribe. And we don't trust people from other tribes. Or to put it another way, us and them. Every time we meet someone new, if we're not extremely mindful to prevent it, our subconscious is busy on our behalf sorting through all the available information, the way someone looks, the way they dress, their body language, the way they speak, everything. All to decide if someone is 'us' or not.

It led me onto an Ancient Greek word (because nothing is new after all, and the Ancient Greeks were more insightful than most), Homophily. This is a word that describes the all too familiar tendency that people show to associate themselves with people who are similar to themselves in significant ways. The phrase 'Birds of a feather flock together' is a direct interpretation of this rather tiresome and unimaginative aspect of human nature.

Have you ever noticed how all your friends are around the same age as you? Groupings of human beings around the world can be distinctly identified by their age, also their race and ethnicity, even earning similar wages for doing similar jobs. These groupings often believe in the same deities also, or absence of them.

It's incredibly difficult to change this mindset. After all, it's hard-wired into our psyche alongside our other irresistible motivations to pair up, settle down and fire some more human beings out into the world. The motivations go hand in hand too – how better to protect our little snowflakes as they grow up into loyal card-carrying contributors to our consumerist culture than to be surrounded with people who think like us, who act like us. And more uncomfortably, look like us.

The monster that is social media has a part to play here. Every time you login to a certain blue logoed social network run by an individual with a lot in common with Mr Data from Star Trek, you're exposed to articles and thinking you might be interested in, based on articles and thinking you have interacted with before. On the face of it this reinforcement might seem like a useful artefact of the design to make it more helpful, but the truth is anything but. Your own quiet corner of your favourite social network becomes an echo chamber, filled with the things you are comfortable with, and none of the things you are not. In this way we can share a tool with individuals who think nothing like us, who believe polar opposites, and never once run across their (from our perspective) puerile ravings.

Is it possible to change our inbuilt homophilic behaviour? Absolutely, but it requires concentration and vigilance, as do all changes to our inbuilt Wiring. And it takes time. The capital city of Italy was not assembled in 24 hours, so the phrase doesn't go.

I've always been a fan of the phrase, 'familiarity breeds contempt', at first because it sounded kind of dark and edgy, but when I really thought about it, I realised I like the way it sounds when I say it to others. I enjoy the faint disappointment in their eyes as I no doubt slide down a Homophily Scale they had been busy constructing in their heads as I was being assessed for integration into their tribe. It also explains the exhilaration that comes with hauling one's carcass out of one's comfort zone.

We crave the familiar, the safe, the routine. There's nothing wrong with that, but then, maybe there's everything wrong with that. If you want to step outside this bubble of contentment and conformity you need to get comfortable with feeling uncomfortable. Look back on your life. How many experiences started with the thought 'I wish I didn't have to do this' and ended with 'I'm really glad I did that'. I'm proud to look back and see more opportunities taken rather than not. It starts with identifying the cues that surround a decision to play it safe, then taking a few deep breaths before doing the exact opposite to your immediate safe impulse.

Scaling up our thinking for a moment, at the national level tribalism is extremely, extremely bad. Those on the other side of the border are bad, while on our side of the border they are good. Dangerous maniacs like the Lunatic in the Whitehouse will have you believe that on the other side of that wall are a bunch of dangerous, narcotic-taking, gun-wielding rapists who want nothing more than to steal your way of life. The only thing scarier than his insane rhetoric is listening to the baying hordes of people chanting along, hanging off his every word.

What is a country? An inherited group of people whose ancestors grew up on the same piece of land as you, which somehow confers some nebulous sovereign right for that to continue in the future. The resources on that land, and under it belong to you and yours. And you'll be damned if anyone is going to take them away.

A great, if horrific example of tribalism played out on the world stage is the guerrilla war that took place during the latter half of the 20th century in Northern Ireland. The conflict went by the charmingly understated name of 'The Troubles', with paramilitary forces determined on one side that the only right answer was to be Irish and Catholic, whereas the other side were convinced that to be British (and for their paramilitaries, Protestant) was the one true way forward. This frankly minor (in the grand scheme of things) distinction led to more than thirty years of violence and bloodshed where considerably more innocent civilians were killed or maimed than among the dedicated participants on both sides.

The European Migrant Crisis is another slightly more recent example. Starting in 2013, increased numbers of people started arriving in the European Union from Africa and the Middle East. Often travelling on foot, with nothing more than the shirt on their back, these poor souls are seeking a better life than the one they were born with. Climate change is a strongly causative factor here, once it becomes too hot, once water is too hard to come by, crops are going to stop obligingly feeding the masses. Naturally, in the newspapers of the western world, in the drinking establishments where those with little to say recite the opinions of the media as their own, this has been seen as a bad thing. They're coming here to take what is ours.

Sigh.

In 2018 a report published in The Lancet, the oldest and most respected medical publication in the world described something rather remarkable. There has been a global decline in the number of children being born. In 1950, women globally produced an average of 4.7 children in their lifetime. I should point out this was a long time before good television like Game of Thrones could fill up our evenings. In any case, by 2017 the fertility rate had almost halved to 2.4 children per woman.

For someone in my situation, observing the decline of the human race from my metaphysical deck chair (with cup holder), with vasectomy under my belt (geographically as well as symbolically), this made little difference to my outlook, but I started to wonder whether others were starting to slowly, painfully come to some of the same conclusions as I did.

Underlying this decline is an inescapable fact. More and more couples are deciding against having children. Their reasons range from a concern about the overpopulated planet of ours, to a concern about being able to afford to go on a nice holiday every year to somewhere where cocktails come in coconuts. Even in my limited social circles, more and more women are reaching middle age without the need to attach a small person to their bosom several times a day. Good for them. I firmly believe the ultimate expression of empowerment is to do what you want, and not what society dictates you want.

I read recently (early 2019) of a new concept called Birthstrikers. No, this isn't about slapping a baby, it's about a global movement of primarily women that recognise the oncoming storm of catastrophic climate change and opt not to put a child through that unpleasantness. Stories like this in the news give me a deep sense of comfort, as they tell me that at least it's not me taking crazy pills when I look at the world and what is about to happen.

All of which means, on a planet where fertility rates have halved in the last half century, where some countries are actually experiencing a net decline in population, and many more a massive increase in the population of retirees (putting an enormous burden on their economy), at a national level how is immigration somehow being still perceived as a bad thing? Grab those immigrants while you can! Honestly... if I felt the average reader of the Daily Mail could understand economics, I'd seriously be tempted to try explaining it.

So here we are. You, me and seven and a half billion of our close chums, trapped down a gravity well in an unremarkable planet in the Orion arm of the Milky Way (not the unfashionable end of the western spiral arm as Douglas Adams once famously claimed). We didn't choose to be here, but here we are, and if we're smart, we're going to make the best of it and have a few laughs while our DNA telomeres steadily shorten as we get older and older. If we're not smart about it, well, that's fine too. I make no judgements, but I'm very happy with an 'us' and 'them' distinction to differentiate us in this scenario.

And if that's where we are, what's to come?

I've already covered a declining population, that's only going to get worse. Maybe we're seeing the hand of Gaia at work, the planetary organism trying to heal itself, though I am more of the opinion that Gaia's solution to its infestation will be to bring in an exterminator in the form of a kilometre sized rock. You may have read the excellent PD James book, 'The Children of Men', or seen the equally excellent film that was made from it. I've often wondered how a person in my position would have reacted to a sudden catastrophic drop in fertility rates. I will likely never know, but it is my fond hope that it would be a quizzical expression and slight shrug of the shoulders before continuing to go about my daily amusements.

As resources start to run out, starting with fossil fuels and ending with water, no doubt we'll experience a further regression towards our tribalist roots. 'Us and them' becomes 'They have and we want'. The dwindling militaries of the world will fight their final conflicts on behalf of the super elite over the last oil fields, then the last arable land, then the last fresh water supplies. While that goes on, we'll be protecting what we do have, building bigger and bigger walls to keep desperate, less unfortunate human beings out.

Improvements in medicine and general quality of life have had an unintended impact on our tribal future. The world average life expectancy in 1900 was just 31, whereas little more than a century later it was over 70.

It doesn't take a mathematical genius to uncover that the years you are alive start from your birth, so when life expectancy doubles, you end up with a higher spread statistically of older people, not to mention more people in general as they have more time to procreate, sometimes starting two or even three families in their lifetimes.

All around the world, we're experiencing the impact of an ageing population, and with it, divisions between the old and the young become stark and bitter. This leads to a democratic bias where older people will vote for initiatives that give them more money or otherwise improve quality of life, making the system of politics easy to game. Say you are an unscrupulous politician, with controversial ideas that make your getting into power rather unlikely. So, you tack on some other policies, like increasing pensions or reducing fuel taxes then suddenly, you find yourself the belle of the silver ball, free to wreak all kinds of havoc for your own short-sighted ends. I have of course in a roundabout way just described how the UK voted to exit from the EU in 2016, but I shall say no more about that, it's a whole other book in itself.

As well as dwindling resources, bitter armed conflicts and the advanced in years taking on their youthful counterparts, I fully expect we'll see further failures of democracy. Civil unrest will lead to the fall of governments, and in their place, I expect more populist rulers will rise, possibly even some dictators. Indeed, while it's too late now, a planetary dictatorship that castrated big business and prevented it from destroying the ecosphere wouldn't have been the worst thing in the world for a few generations. Well, maybe it would have literally been the worst thing in the world, but history (from a standpoint where we'd survived into the future to have a history) would look upon certain aspects of it as useful, if not essential.

Try as we might though, we can't escape our primitive tribal roots. Wherever you find human beings, you will find tribes.

What else could the future hold in for our tribe? I personally still harbour the deep, uneasy suspicion that reality as we know it is a construct, a simulation in an unimaginably powerful computer, in an unbelievably distant future.

Does it have a plan for us? Is the future within the construct predetermined? How many iterations have you and I, as unwitting characters in this absurd telenovela been through already? Is there a purpose, is the construct seeking something beyond enlightenment and understanding? How much free will could you have under such a system? Go and get a cup of coffee. Or don't. Did the system know what you were going to do before you did it?

The origin of this paranoia that not everything is as it seems, is older than you think. It was first postulated by philosopher René Descartes in 1641 in his book 'Meditations On First Philosophy'. He states: 'there are no certain indications by which we may clearly distinguish wakefulness from sleep… it is possible that I am dreaming right now and that all of my perceptions are false'.

More recently Swedish philosopher Nick Bostrom has been putting his little grey cells to work on the conundrum. In his 2003 paper 'Are you living in a computer simulation?' he argues that at least one of the following statements is true:

(1) The human species is very likely to go extinct before reaching a 'posthuman' stage
(2) Any posthuman civilization is extremely unlikely to run a significant number of simulations of their evolutionary history (or variations thereof)
(3) We are almost certainly living in a computer simulation

It all comes down to your confidence in the future. I've been expending thousands of words now in my belief that (1) is true after all. Perhaps you don't believe me though, maybe you're less cynical –

which is statistically highly likely on the face of it. Let's say you are convinced that mankind has what it takes to pull a rabbit out of the hat and make it through to a golden age future where all things become possible, then (3) becomes the most likely of the above to be true. The paper is well worth a closer read anyway if you're interested, I'm not going to copy it verbatim here but it's freely available online, a more interesting 5712 words you will be hard pressed to find.

Having given you that get out clause, I have an alternative, perhaps even more outlandish theory behind the simulation to present to you, a theory that allows one both statement (1) and statement (3) to both be true.

I don't think it's being run by our descendants deep downstream. Don't get me wrong, it's definitely run in the far future – a quantum computer from the present day is a mere speak and spell compared to the machine that would be needed to run our entire reality. I would conservatively estimate we need trillions of times more processing power than is currently available to our civilisation.

No, I don't think it's run by humans at all. After all, it's typical anthropocentric thinking to put humans at the centre of everything, when the truth is anything but. It isn't that long since we thought our planet was the centre of the universe after all. I am of course, talking about aliens, creatures who evolved somewhere else, billions of years from now. They are running the show, and grudgingly paying the exorbitant electricity bills our simulation is doubtless running up.

Perhaps those future creatures discovered the dusty ruins of a civilisation far out in one of the milky way's spiral arms, orbiting a red giant that billions of years earlier had been a yellow main sequence G type star. Perhaps they comprehended from radioactivity or atmospheric toxicity that the end was our own doing. Perhaps they just want to know why, how a sentient race that achieved so much in such a short space of time let this happen to themselves.

In that scenario, they may have been insatiably, even morbidly curious they might glean what they can from records that still existed, from painstaking archaeological research, and spin up a plausible present day (from our point of view) scenario complete with billions of procedurally generated entities. Each entity would be capable of independent thought, and the creators would have them run around the scenario, again and again like mice in a maze. All the while, the mice are blissfully unaware that the maze exists.

Many respected scientists, including Neil deGrasse Tyson and Elon Musk believe that not only could our reality be a figment of a quantum computers imagination, they consider it actually more likely a simulation than not. Musk even went so far as to say the changes of this not being a simulation are one in billions.

Something of a rebuttal came in the form of the results of experiments conducted in 2018 at Oxford University by Zohar Ringel and Dmitry Kovrizhi. They used a technique called Quantum Monte Carlo to develop a simulation of their own, of just a few particles, and found that the processing requirements don't just increase when you simulate more particles, the increase is exponential. They postulated that to accurately simulate, down to the quantum level, all the information for just a few hundred electrons, they would need conventional computer memory that contained as a storage medium, more atoms than exist in the universe.

So, can we rest easy then? What we see out of the window (the Sydney Opera house as I type this) is really there? If I were to go outside and walk to it, could I lay my hand on it? Only joking, or at least, I rather doubt it. Firstly, we're assuming the far future curators of a simulated reality are relying on conventional computer memory, are we? That they queue up at Walmart to buy hard drives to store the known universe on? Even disregarding that, a more fundamental problem exists; when the only tools to measure reality exist inside reality, how can you ever be sure they aren't showing you what you expect to see, rather than what is there?

There's the rub. It's a great intellectual experiment, but nobody can prove it. On the bright side, nobody can disprove it. I could tell you about phenomena such as running into someone right after you've been thinking about them, déjà vu, the feeling of falling as you fall asleep fall under the category of rendering errors, but that would be fanciful, even for me.

Say it's true, what then? Economist Robert Hanson had a dark take on it, that being self-aware within a simulation could lead to a lack of empathy, and disinterest in planning ahead – live in the moment because it's the only part of your life that's even vaguely real. He said: 'your motivation to save for retirement, or to help the poor in Ethiopia, might be muted by realizing that in your simulation, you will never retire and there is no Ethiopia.'.

No, for me Elon Musk has it right. He's convinced we live in a shared simulation yes, but does it stop him achieving, stop him striving for the stars? No, of course it doesn't. Okay, it's possible that reality might be false, but it's still the only one we got, and as Groucho Marx famously said: 'I'm not crazy about reality, but it's the only place to get a decent meal'.

The best evidence I personally have for this being a simulation is the stock market, and specifically, the patterns that traders look for in the insane stream of data that accompanies the flurry of buying and selling activity. One of the main techniques used is called a Fibonacci Retracement, and is based on the work of Italian Mathematician Leonardo Fibonacci in the 13th century. It was about here my eyebrows began to narrow, what on Earth does 800-year-old math have to do with the most complex system that human beings have managed to accidentally create?

I won't get into it here, it deserves its own book, but the Fibonacci numbers are as close to the source code of the universe as we are likely to get. They occur, everywhere and in everything, throughout nature. The numbers of Fibonacci are best expressed as a spiral, and the ratio of this spiral can be seen everywhere from a snail shell to the Milky Way spiral galaxy we inhabit. From the infinitesimally small all the way up to the very top, you see the Fibonacci sequence at play.

The same is true of the stock market, an anarchic system of ones and zeroes that barely follows logic, let alone predictability. Yet, I have it on very good authority from friends who indulge in day trading that the Fibonacci numbers are key to understanding retracements, stock prices reversing their direction, going from low to high or high to low.

The conventional wisdom, or at least flimsy explanation is that since the Fibonacci sequence expresses itself everywhere in nature, so does it express itself in the creations of natural beings. That we have subconsciously somehow imbued the stock market, a system so sophisticated I'm surprised it's not ordering its own coffee, with natures' own Fibonacci numbers.

I don't buy it. My personal theory, since you asked, is that the Fibonacci numbers are a glimpse at the operating system that underlies the simulation, cracks in the façade that envelopes our consciousnesses. They occur throughout our world with such regularity that a quarterly publication exists to document their occurrences. I'm seriously considering subscribing.

Perhaps the people who believe that the Fibonacci exists in the stock market are half right. Perhaps as the stock market is a simulation within a simulation the inherent errors are compounded (or squared) to the point where we can see them, in the same way that the only way we could hope to view a four dimensional cube (aka hypercube) would be to observe its shadow.

Another amusing and only slightly sardonic observation of simulation evidence is the existence of the Durian fruit. Originating from South East Asia, this fruit variety provokes an intense reaction in everyone who smells its delicate aroma. The interesting thing is nobody can agree on the smell. To one person it's the most delightfully mouth-wateringly sweet thing they've ever smelt, yet another person will note with extreme its remarkable aromatic similarity to raw sewage. Conventional wisdom will hold that something fundamental in our olfactory system has branched during our evolution, producing such wildly different interpretations of the same smell. Unconventional wisdom? Glitch in the matrix.

I would also like to note that, if this truly is a simulation, the fact that everything we do that is fun (with the notable exception of orgasms) tends to shorten your existence within it is a rather sick joke on behalf of the programmers. And if nothing else, the world being a simulation does answer one very important philosophical question.

If a tree falls in the woods, and no one is around to hear it, does it make a sound?

No, because it would be a waste of processing power.

# CHAPTER 16 - NOT WITH A BANG, BUT A WHIMPER

You may be familiar with the old phrase the chapter title refers to, which the final line of a poem, 'The Hollow Men' by T.S. Eliot. The poem was written in 1925 in the aftermath of World War One, but its bleak tones have led to it being appropriated by anti-nuclear factions from the 1950s onwards. I'm going to appropriate it here for a tangential use around the utopian holy grail of clean free energy: fusion, and in particular my own rather sceptical vision for its future potential, complete with some of my trademark realistic reasoning.

Before we consider Clean Energy, let us take a quick look over the alternative, the appropriately named 'Dirty Energy'. Simply put, it's any energy, that in generating to power our toaster ovens and margarita machines, harms the environment in the form of excess heat or pollutants sent into the ecosystem such as carbon or CFCs. I'll make the definition even easier for you – if it comes out of the ground and needs to be burned, harming the delicate little bubble of air we like to use for breathing, big tick in the dirty energy category.

That accounts for coal, oil and gas, the big three. Nuclear fuels have to get an honourable mention also for their impressive amount of unbelievably toxic and radioactive by-product – and the fact that these too will run out, at the present rate of consumption by around 2150. We're already pretty much out of plutonium, used in large part by NASA for deep space probes such as Voyager and Pioneer, quietly launched into orbit a few pounds at a time over the past few decades – a fact I found not scary at all. Amazingly, and equally short-sighted, a couple of wood fuelled power stations have now started cropping up. It makes a twisted kind of sense, if we're running out of the processed hydrocarbons in the ground, we might as well start using the unprocessed kind. It's not like trees have anything to do with maintaining our ecosystem is it?

How about bioenergy? Simply put, this is the energy that is contained within biological matter, or biomass for short. There are a great many possible sources of bioenergy: corn, sugar cane, rape seed, straw. How do we get at that energy locked up within it then? Simple, we do the same as we would when we hang around other teenagers when we're 14, we start playing with matches. Or perhaps that was just me. Bioenergy isn't going to be the solution to our energy woes, since this form of supposedly clean energy still emits carbon into the atmosphere since we're still setting stuff on fire and brings with it a whole host of new environmental and social problems.

Social problems, you say? Well, to start with, when farmland is given over to bioenergy production, great for the farmer, there are some decent incentives starting to appear for bioenergy production. But what was growing on that land before it became a source of energy to our civilisation? A source of food for our civilisation. So unfortunately, we're robbing Peter to pay Paul.

Out of curiosity I looked into the origins of that saying, I suspected it originated as a biblical reference, and it does, but not in the way you might think, no disciples were held up at gunpoint. Legend has it, it's actually a tax-based quote (as many of the best always are) and is related to prioritising the payment of church taxes in Britain for St Pauls Cathedral in London rather than sending it to St Peters Basilica in Rome. Ah tax, you sexy, sexy bitch.

Another issue of bioenergy production is that converting land to its production often results in additional deforestation. I really would like to see a monetised way of keeping trees on your land that doesn't involve eventually chopping them all down to feed into a lumber mill. Were it financially viable and the world a fair place, we'd call them 'oxygen farmers', and it would be the most noble job in the land. Someone please, go figure that out and come back to me, would you?

Or what about Waste-to-Energy? Try waste of time. This is the slightly inane (or maybe that should be insane) idea that we should be burning all our household waste in order to generate energy from it. Putting aside the ridiculous amount of pollution this gives off into the atmosphere, on a commercial level it's actually been found that it takes more energy to burn household waste than you get back from it and is massively less efficient than recycling.

As we covered in our Hydrocarbon Addiction chapter, major world powers, including the USA and China remain fully behind fossil fuels in a big, and largely unstoppable way. How long will the party go on for though, that is the big question that sets a ticking time limit on us needing abundant cheap and clean energy to preserve the status quo. British Petroleum has been uncharacteristically open on the subject, at current rates of consumption (which is set to rise as the developing world gets its first taste of western standards) our supply of crude oil for example will last for around 53 years. Or to measure it another way you might be more comfortable with, 6.6 Game of Thrones (the exchange rate is that one Game of Thrones = 8 years). That's not that long.

Clean energy on the other is very easy to recognise. It's renewable, so we don't run out of it anytime soon, and it doesn't do irreparable harm to the ecosystem we depend on to live, breathe, eat and sleep. Before we come to the main course of our chapter let us explore a few of the renewable energy sources.

Solar Power has surged to prominence over the past 40 or so years, and takes many forms, but the one we are most preoccupied with is the conversion of sunlight into an electric current using photovoltaic cells. Don't get me wrong, solar energy is great, if you read my last Opus you'd know that my retirement plan includes a solar farm in a corner of the world remote enough to ignore all the problems we've brought on ourselves for the rest of mine, and my closest friends natural lives. My concern with solar energy being a solution to the woes of our entire species sits primarily with what we make the solar cells out of. Hydrocarbons. The dinosaur bones that could be our long-term saviour, are our short-term dirty energy source. Once we're out of hydrocarbons, no more solar cells. In addition, the fossil fuels lobby is brutally against solar energy research, seeking to maintain their monopoly while they still can, which hampers research and development efforts to improve solar efficiency.

Another fun contender for clean energy is geothermal energy, harnessing the power of the core of our planet (which ticks over at just under 4,000 degrees Celsius) in order to produce electricity. Geothermal comes in two forms, heat exchange, which can be great to heat homes, and to generate electricity. I've first-hand witnessed the heat exchange type, at a friends' place in Switzerland and while extremely impressive, the primary downside was immediately obvious. It's really, really expensive. Other downsides include greenhouse emissions, geological instability (as with fracking it's invasive as requires pipes to be drilled) and worst of all, location, location, location. You can only build a geothermal power station at very specific locations on the planet, at the sites of geothermal reservoirs near enough to the surface to be utilised. There aren't nearly enough reservoir locations to make a significant dent in the energy requirements of our ever-increasing planetary population.

Or what about wind power? I make a regular train journey through England, and pass several wind farms, and I never fail to marvel at the enormous, quiet majesty of the wind turbines standing tall over the green landscape. It's extremely clean and renewable of course, but as with solar, leaves you somewhat at the mercy of the elements. If there is no wind blowing, there is no power to generate. For this reason, most wind farms operate on about 30 percent of maximum capacity. This dramatically limits the locations that can be selected for wind farms, which again limits the potential gain. Some wind farms have also required extensive clearing of forest land in order to build them, which rather defeats the object.

How about Tidal Energy? While it is much more reliable than wind and solar, since the tides can be accurately predicted, it still suffers from siting limitations, as only very specific areas offshore can be converted into tidal energy farms. Also, the way the tidal surges occur, only about 10 hours per day can generate usable energy. Hydroelectric dams also technically fall under this category but have massively declined in use over the past half-century, except in China, where they regularly displace millions of human beings to create ever more floodlands, all in the name of keeping the lights on.

On the 23rd of March 1989 the most interesting thing to ever happen to our species almost happened. It all started when scientists Martin Fleischmann and Stanley Pons issued a press release. They had been playing around with the electrolysis of heavy water on the surface of a palladium electrode, and reported that this reaction produced excess heat, as well as trace amounts of neutrons and tritium. Since these were known by-products of nuclear fusion, the scientific community, naturally enough, lost their shit, collectively speaking.

Nuclear fusion at room temperatures, known more commonly as:

Cold fusion.

On the face of it, this phenomenal discovery was world-changing, with the potential to be the greatest economic upheaval since the Pennsylvania Oil Rush. Think about it, heavy water is easy enough to produce, and we've got lots and lots of palladium. If we could smash these two things together and produce energy, so to speak, then a lot of our environmental problems would be over before they began. I certainly wouldn't be sat on this train writing this book. I might still be sat on the train, though it wouldn't be powered by diesel.

Fleischmann and Pons were thrust into the limelight, and for six months scientific teams around the world got to work on checking their homework, and that was where it all went wrong. Not a single lab anywhere in the world could replicate their results and produce more energy than was being put into the process via the electrodes. Within six months the cold fusion claims were firmly debunked and thrown on the scrapheap of science along with Einstein's Cosmological Constant, Fred Hoyle's steady state model of the universe, and that time in 1862 when Lord Kelvin announced that Earth was only 20 million years old.

What on Earth possessed these two leading scientists to make such claims? To be fair, it wasn't a deliberate hoax, but hubris born of many experimental errors that came to light when replicating the findings. Their reputations were shattered, never to recover, and the fabled 'star in a jar' was not to be.

Can you imagine what the world would be like today if that experiment in 1989 had proved successful? The fossil fuel industries would have gone into immediate decline (boo hoo). I can well imagine the utter shock in the expensive boardrooms of such companies when that press release came out, that would have been priceless to witness. The planetary ecosphere would be well into recovery mode by now. With fusion powered rockets, we'd have a permanent presence on Moon, probably Mars as well. We'd even have those flying cars now that Robert Zemeckis promised us back in 1985.

Fusion is perfectly and scientifically possible right now, it's just hot rather than cold. It actually happens in nature all the time, everywhere you look in the sky. It's the mechanism by which stars shine, but it requires extreme temperatures, hence why stars do it so well and human beings in labs don't.

Fusion is the technical name for two hydrogen atoms combining to form an atom of helium. The waste, leftover mass of that combination (or fusion) is transformed into energy. Boom, quite literally. It's that simple too, I even managed to explain it in a few sentences. And like most of the simplest things, it's astronomically difficult to put into practice.

To make fusion work by conventionally understood methods, the hydrogen atoms have to be heated up just a little (to around 100 million degrees), which turns the atoms into a high energy plasma, which then must be confined together for long enough for the fusion to occur. In stars this is simplicity itself, gravity is on their side acting as the catalyst. On Earth, with gravity set at good old 1G (9.80665 newtons of force per kg) not so much. The gravity we know and love is nowhere near enough to compress high energy atoms together in order for them to fuse.

On Earth, the best ideas for non-gravitational confinement of high energy hydrogen plasma revolve around magnetic confinement, since the atoms can be held in place and confined while they are heated, before they fuse, releasing their energy. Discounting the ridiculous safety concerns of working on a commercial scale with a material that has a 100-million-degree temperature, there is another more practical problem. While fusing hydrogen into helium on a commercial scale has the potential to release an enormous amount of energy, it also takes an enormous amount of energy to heat up and hold the hydrogen plasma in place, so long reaction times are necessary to make it viable. Not only that, but a by-product of this type of fusion production is neutrons, which tend to make the housing of the fusion reactor rather radioactive.

An alternative to magnetic confinement that shows some promise is inertial confinement, which is rather less Star Trek and more Start Running. In this approach, a small pellet of frozen hydrogen (to freeze a gas we're talking insanely cold too) is hit with a laser beam, the frozen nature of the hydrogen means that the atoms fuse before they have time to fly apart.

Whatever the fusion process, nobody is quite sure whether the maths works in our favour to ever make hot fusion commercially viable, hence the giddy excitement back in 1989 when we briefly thought someone had cracked it without the need for the superheating element. For a very brief moment, we thought our energy and environmental problems were over. Knowing us creative old humans, we'd soon have replaced it with something equally insane, but it's nice to imagine.

Since the ingredients to the Fleischmann/Pons Cold Fusion experiment are a vacuum flask, a cube of palladium, some deuterium and electricity, I decided last night that I'm going to repeat the experiment myself. It's possible that perhaps, just perhaps there is a variable that nobody is seeing that meant on that fateful day in 1989, the future for one moment became possible.

I'm no scientist, certainly not by trade, and if anything, my skillset resides more in the big picture world, what might be called futurism if

I was feeling fanciful (and when am I not?) which at a stretch could be described as 'seeing what others haven't seen'. For that reason, I'm inclined to give this a go for my next project and attempt to replicate the Fleischmann/Pons Cold Fusion experiment.

As with all my entrepreneurial efforts (of which there are many, as who likes having free time) my special skill in this endeavour will be knowing some people. Over my many years of bouncing around industries and the world, I have built up an extensive cadre of like-minded souls, people whose personalities, experiences and abilities complement my own in all kinds of innovative ways. You know who you are, stand by your phones.

I'm curious to see if my reasonably abnormal grey cells can conceive of a factor for successful room temperature fusion that hasn't been attempted before, which while that might sound arrogant is also perfectly sound scientific process. The factor for example, could be something to do with geographical location, geological composition of the surrounding area or time of year / position around sun at the time of experiment. These diverse factors all might relate to the possible existence of subtle underlying magnetic fields which have some causative factor on whether fusion occurs. Or maybe it's Ley Lines – that would amuse me no end.

I'm also curious to explore whether applying harmonic fluctuations to the electrical field you pass over the palladium have any impact at all. Harmonic vibration is an underlying and poorly understood concept in the universe. Our universe according to research vibrates at a frequency of $2.2963 \times 10^{-18}$ Hz, and perhaps some manipulation of this type is necessary in order to make the magic happen. After all, as any astronomer will tell you, the universe is music.

The origin of humanities exploration of the physics of sunshine goes back to the early days of the 20th century. It was astrophysicist Arthur Eddington who first suggested in 1926 that the energy source for stars was the conversion of hydrogen into helium, and was at the time ground-breaking, and laid the foundations for the way we understand the universe today.

The first demonstration of fusion in a lab was actually not that long afterwards, in 1934 by Ernest Rutherford, who was able to demonstrate the fusion of deuterium (heavy hydrogen) into helium, producing what he termed 'an enormous effect'. He used an early particle accelerator to shoot the deuterium into a various metal foils coated with different atoms, comparing the energy levels needed for the different reactions and ultimately realising the deuterium-deuterium was an oddly interesting combination.

His assistant, Mark Oliphant went on to refine the experiments further, discovering tritium (the radioactive form of hydrogen) and Helium 3 in the process, both of which are widely expected to be key ingredients in a stable productionised fusion reactor. Let's keep using the word reactor by all means, but knowing the human race as we do, free to interchange it with 'bomb' as needed.

The first patent for a fusion reactor, a version of Rutherford and Oliphants work that produced more power than went in and was able to capture said power for practical applications was submitted in the UK in 1946, though it took until 1951 for the next leap in fusion evolution.

First in March 1951, Argentina falsely claimed they had achieved controlled thermonuclear fusion, a claim that was soon revealed to be a prank. A prank?? Oh, Argentina you trickster, when will you stop joking around and take world affairs seriously? Then two months later in May, Lyman Spitzer, noted astrophysicist, proposed the Stellarator Fusion reactor design. While a great name, it was not a successful design, and many attempts through the fifties and sixties to create practical reactors from this complex geometric design failed.

This design was succeeded by the 'toroidal' (doughnut-shaped to you and me) design known as the Tokamak. Originally conceived in the 1950s by Russian scientists Igor Tamm and Andrei Sakharov, the Toroidalnaya Kamera i Magnitnaya Katushka (Toroidal Chamber and Magnetic Coil), became abbreviated to the 'Tokomak', which turned out to be considerably easier to say after a few vodkas. Throughout the sixties reactors using the Tokamak design reported better performance than their Stellarator counterparts but were repeatedly dismissed due to potential problems in the way temperatures and reaction rates were measured. It eventually took a British delegation to Russia in 1969 to confirm the findings, and the Stellarator was left in the dust.

There are three key performance factors that need to be considered in creating a workable fusion reactor which are, in no particular order: plasma density, reaction temperature and confinement duration.

In the nineties, the Joint European Torus (JET) in the UK and the Tokamak Fusion Test Reactor (TFTR) in Princeton, USA made significant strides in the area of fusion reactor technology. Since then, France have joined the party with the Tore Supra Tokamak which attained a record-breaking fusion duration, which remains a continued obstacle for a production-ready fusion reactor design. Meanwhile the JT-60 Tokamak in Japan has managed the highest values of all three factors combined, and the Princeton TFTR managed a whopping temperature reaction into the hundreds of millions of Celsius. Steady on Princeton, you don't want to lose your eyebrows... or campus buildings.

The holy grail of fusion development is a device which surpasses the 'break-even' point, where more energy is released than is put in to start the reaction. The longer the reaction times, the closer we creep towards this utopia. The JET, TFTR and JT-60 are all getting within spitting distance of this goal.

Which brings me onto ITER.

ITER was first proposed in 1985 by General Secretary Gorbachev of the (at the time) Soviet Union to US President Reagan. To this day it remains a classic example of nationalistic bickering triumphing over common sense over what could well amount to the survival of the species. It started with an agreement by the delegates of the Geneva Superpower Summit to collaborate on a large, internationally developed, internationally owned fusion research facility that might one day solve all of the world's energy problems.

Let me paint you a quick mental picture. It was 1985. The Compact Disc had just been launched. French secret agents sank the

Greenpeace ship Rainbow Warrior. Back to the Future came out at the cinema. Coca Cola still thought that New Coke was going to be a really good idea. The pop music industry unites to perform 'We are the World' for the starving children of Ethiopia, as if they weren't already suffering enough.

Even in those carefree days a person with just half an eye on the future of the species would wince about the upcoming collision of energy and environmental concerns. From Gorbachev's radical proposal to Reagan, it only took a year to finalise the agreement to move forward, which for government bureaucracy must surely be some kind of record.

From there the conceptual design work would race by in just 14 short years, barely enough time to conceive a child and have it become a teenager and tell you how it ruined his/her life (to which you would try not to reciprocate the sentiment).

I actually met someone from the ITER project team back in 2000, a design engineer who worked out of a facility in Munich. He was the father of a friend of mine who I made the acquaintance of whilst doing what most 23-year-old males do if they grow up in Europe, heading to Oktoberfest. While gamely tolerating our late-adolescent foolishness, stumbling back in at all hours full of beer and pretzels, he talked of the lofty goals, of the immense torus shapes that were being conceived. He also told me freely what a gravy train the ITER project was considered to be, and of the many, many people working on it who felt the same as he did. I recall telling him at the time how cool I thought the project he was working on was, he shrugged with a smile that I now understand was pitying.

Following the design phase, the timewasting really kicked in. It took just five years to argue about the best location to build the reactor, everyone of course wanted it in their backyard for all the street-cred it would do. The final two contenders in the geographical race were Japan and France. Japan argued it had the technical edge, while France argued it had much better cheese (and more open space to build such a facility). Finally, painfully, a compromise was reached, the R&D facility building the toroidal field coil magnets needed for the device would be manufactured in Japan, while the final reactor site would be located at Aix-en-Provence in Southern France.

Since each toroidal field magnet (of which there are eighteen) measure approximately 9 x 17 metres and weigh in at 310 tons, transporting them the 10,000 kilometres from Japan to France will be something of a massive challenge by itself. The world's largest cargo aircraft, the Soviet-era (but still flying) Antonov AN-225 Mriya can only carry 250 tons, and has a cargo width restriction of 6.4 metres, so that won't work. In simpler times we'd no doubt have used felled trees to form some kind of sled in the same way the druids moved the stones of Stonehenge more than 200 kilometres from the Presili Mountains to their final standing place. With a bit of luck though saner minds will prevail and cargo ships will transport the magnets, though no doubt the procurement paperwork for a new super-heavy plane sits on some bureaucrat's desk somewhere itching for rubber stamping.

Meanwhile, at the reactor site in the South of France, ground was finally broken in 2005, a mere 20 years after it was originally proposed, barely enough time to plant an acorn and have it grow into a mighty 30-metre-high oak tree. From there on in, development raced along at breakneck speed. Just 13 more years passed until the first component was put into place in the large building that had sprung up seemingly overnight. The facility is expected to go into partial operation by 2025, and become fully online by 2035, just 40 years after its conception, which hardly seems like enough time to grow up, have a career, become bored of it and craft books that nobody will ever read...

Some facts about the ITER, just to put it into some kind of megaproject context for you. When completed the reactor will weigh 23,000 tons, consist of more than one million components and have the capacity to contain a plasma volume of 840 cubic meters, more than a 20-meter-long swimming pool.

The stated objective of ITER, just as Gorbachev had in mind is to demonstrate the technical feasibility of using fusion energy to generate power. It is planned to conduct extensive and sustained burns of deuterium-tritium plasmas with a goal to release ten times as much energy as is input for the initial reaction.

Assuming ITER completes its objective successfully and power generation using fusion is proved possible (and to some extent safe) its successor, the DEMO plant (imaginatively short for demonstration) is expected to go online by 2033 and run until 2050, though let's call it 2060 to be on the safe side in case there is a protracted argument about wallpaper. For DEMO to be declared successful it needs to produce 2 Gigawatts of power continuously, which is significantly greater than the current generation of coal power stations, for an input of power no more than 4 percent (80 million watts) of the total generated.

Once this is completed, the way is paved to build out more fusion reactors around the world, with every member nation freely sharing in the designs brought about by the vision of Mikhail Gorbachev just 75 years earlier, who assuming he is still with us at that time would be celebrating his 129th birthday when other nations start to plan their reactors. I look forward with great anticipation to the regional arguments by which county or state with a country gets the rights to site the reactors.

Barring a huge amount of scientific progress and hard work, it all sounds feasible right? It will be monumentally difficult, but nothing worthwhile is easy. After all, there's a plan, broken down into small enough sections to sound achievable.

Well, you see, it all comes down something I call the narrowing tunnel.

Since I was very young, I have experienced a recurring nightmare about moving along an upwards tunnel that is progressively getting narrower around me. Sometimes it's a very steep staircase. I can picture it right now narrowing around my shoulders, closing in as I attempt to ascend to another level of some structure I have never been in. Above me somewhere I can sense a much wider open space that I cannot quite reach.

The dream always ends the same way, I wake up gasping for breath, claustrophobia reflex well and truly tripped, and usually have to open a window while my heart slows down from the marathon it thinks I have just run. To this day there is a bar in Amsterdam (Louis' on Damstraat) where I can't go to the toilet because the incredibly narrow (a Dutch architectural idiosyncrasy) and low-ceilinged staircase that curves up to the bathrooms because it triggers the fear I feel in the dream. It also occurs to me now that there is a reason I never became a submariner.

Hopefully you begin to see the glimmer of a point. Our species is busy consuming all available resources at breakneck, reckless speed. You name it, we're using more than can be reasonably replaced. Food, water, energy, every kind of material and manufacturing process. Based on my analysis of fusion, we're realistically speaking 100 years away from this utopia of fusion power at an absolute minimum, which while the fuel source for fusion is reasonably abundant, it's still a highly dangerous and radioactive process. That is fine if you happen to be a star (as in a stellar object rather than a Kardashian sister), but for us poor dumb humans trapped down a gravity well, things are rather more complicated.

My scepticism for fusion being the answer to all our problems boils down to a single question, semi-rhetorical because while neither you or I are qualified to answer it (nor are we fortune tellers), I have a healthy suspicion. The question is this:

'Can we deliver on the promise of (not exactly) clean, cheap and abundant energy to our civilisation before we lose the ability to create it in the first place?'. Think carefully about what the next hundred years of resource shortages might hold before you respond.

I think I know the answer, what about you? This is why the failure to deliver on the Fleischmann/Pons Cold Fusion approach in 1989 was so completely heart-breaking. It could all have been so very different for our species had all that fleeting promise been borne out.

# FINAL MUSINGS

So here we are again. It's a ridiculous paradox on the face of it. Humans as a species busy themselves destroying the ecosystem they utterly depend on, yet individually they will strive to survive with every fibre of their being. The real fighting to survive won't start though until we face the realisation that we are past the point of no return.

You may have heard the (hopefully) allegorical story of the frog and the boiling water. If you throw the frog into water when it's boiling, it jumps right out and, scalding aside and probably very annoyed, he/she can get on with their life. However, throw the frog into water when it's cold, then heat the water slowly and the frog will boil to death, failing to notice the environmental change until it's far, far too late. Disregarding the 'cruelty to animals' motif, there's a lot to learn from this story about the danger of slow change, which I fervently hope is more thought experiment, or at least from a simpler time, rather than a regular feature in biology labs around the world.

Top of the class, dear reader. We are the frogs. The temperature is getting warmer, but we're doing nothing about it. But Tim, I hear you shout (or Mr Biddiscombe if we haven't met and you have an overabundance of formality) we have a much better nervous system than a frog and would notice the temperature change in time. I don't doubt it, would be my response, however individual humans are not the frog, humanity is. Our collective species-level equivalent of a nervous system is numb and slow on the uptake, blind to the changes that reacting to might otherwise save us.

Earlier in the book we talked about the Holocene Extinction, which conveniently for record-keeping started at the beginning of the Holocene Epoch, about 12,000 years back when the glaciers retreated back up to the North and South poles where they belonged. With the glaciers out of the way, our collective ancestor, Mesolithic man (and woman) crawled out of their caves blinking in the sunlight like Phil the Punxsutawney Groundhog. More recently, scientists have proposed the Anthropocene, as a new Epoch, which would effectively terminate the Holocene Epoch a little earlier than scheduled, to be replaced with a new one especially reserved for us, and our rampant disregard for the blue-green mud ball we like to stroll around on top of. More specifically, a new Epoch reserved for the end of us.

Scientists have yet to agree a start date for the Anthropocene, but arguments range from the start of organised agriculture (8,000 years back), to the invention of the first steam engine in 1712 (as a marker indicating the official start of the industrial revolution) and even as recent as the Trinity Test in 1945 where the first nuclear weapon was detonated. An end date for the Anthropocene isn't of course known as it's in the future, but it will mark the end of an ecosystem that can support human life. Nature will of course go on in some form, we won't be able to exterminate 100 percent of all the plant and animal life no matter how hard we try. Some weird tube worms 5 kilometres beneath sea level will happily go on evolving, unknowing and uncaring of the trials and tribulations of the hairless bipeds who made a serious mess of the ecosphere above them.

While the Anthropocene Epoch hasn't yet been approved amongst the geological powers that be, it's nonetheless grinding its way through the endless layers of administration that surround most, if not all our scientific disciplines. In fact, let me lay this out for you. The Anthropocene Working Group (AWG), which is part of the Sub-commission on Quaternary Stratigraphy (SQS), which is in turn part of the International Commission on Stratigraphy (ICS), voted in 2016 to put forward a formal Golden Spike (GSSP) proposal to enshrine the Anthropocene epoch into the Geologic Time Scale (GTS), which presented the recommendation to the International Geological

Congress (IGC). How's that for pointless bureaucracy? And you thought local government was bad. Death by Three Letter Acronym (DTLA for short).

Strictly speaking though, we're only dealing with nomenclature here. This extinction is happening whatever you want to call it. In fact, to paraphrase Neil deGrasse Tyson, go ahead and not believe in it if it helps you sleep at night, neither the upcoming extinction nor myself will care a jot.

Humans will try to fix the planet of course, we're not idiots. Not completely anyway. The issue is the large-scale efforts to save our species will only happen once the politicians and rulers of corporations are inconvenienced. Once their comfortable existences are threatened, perhaps those influencers and decision makers in the 1 percent will finally do the right thing. Rising energy prices? Not their problem, they can still afford electricity, or even if it comes to it, acquire the facilities to generate their own. Maybe the corporations of the next generation will buy up the last of the nuclear reactors to privately power their ailing empires, or on a more regional basis simply stand up acres of solar cells to keep the estates of the rich and shameless lit up like a Christmas tree at night. How about the inevitable shortages of water? Not for the rich. Money will open every door, or faucet. Same for food. Same for everything, in fact.

Surely the people won't put up with that, you wonder musingly. The cynic in me notes that they are already, but nonetheless, I tend to agree with you. Every civilisation is only three meals away from revolution, as history has shown us time and again, the same will be true for water, energy, and even I suspect internet bandwidth. After enough riots, enough massacres in townships in Africa that have been without water for weeks, the 1 percent will start to take notice. Change will come. In the Western World, that will probably involve certain stronger countries annexing certain weaker countries with fresh water, or other resources that we need to preserve our way of life just that little bit longer. Innovations will help us here too – necessity is the still mother of invention and improvements in water filtration, solar power, crop pollination will refill the leaky bucket to some extent. Samuel Clemens, aka Mark Twain once said: 'A sentence of death focuses the mind wonderfully'. He could not have put it better.

Whatever shape and form the efforts take, it will have the depressing distinction of being our last megaproject as a species.

I'm sure I could qualify as a doomsayer, what with all this doom I keep talking about. I'd like to highlight that most of those who fit that category seem to have an agenda or at least be selling an idea about how their personalised doom could be avoided, or at least what to do when it comes. No such luck here, sorry about that. About forty years ago it could have been different, but we were just hitting our capitalist stride, swinging a wrecking ball at nature with as much force as we could possibly wield.

So then, without further ado (or at least only a small amount of ado), what are our options?

1) Do nothing. Draw up a deckchair and pour yourself a margarita while Rome burns. This is the path I am on, if it wasn't clear enough already, but I'm also hoping some of the more contrary among you may consider door number...

2) Do something. I can't for the life of me see anything that would be effective except in soothing whatever personal guilt you may be carrying around. After all, it's not you who personally destroyed the ecosystem congruent with species survival. But they look a lot like you. They had two arms, two legs, one head, carried around 200 odd bones in their bodies. They just had a lot more money, power, privilege, and above all a deathly fear for some reason of one day losing it all. In fact, the one thing you absolutely, definitely possess that they do not is empathy, and isn't a moral victory the most satisfying kind?

Perhaps I'm wrong. It's been known to happen, sometimes several times a day. Maybe you are the person who can make something of option two. After all, sometimes the coin lands on its edge...

What form might that effort take? I'm not the one to ask, since my answers might stray into supervillain territory, though perhaps there is no hypothetical being better qualified to answer, since it's clear that extreme steps are needed. Since you asked, or at least didn't skip over the hypothetical question, let's explore it. Be warned, there is going to be nothing humane about what I have to say, but I can live with that, since in some respects I long ago ceased to be a card-carrying member of humanity. As I once remarked in a previous book which I may have mentioned, 'I might look like you, but I am not you'.

Don't get me wrong, I love humans, some of my best friends are human (I hope my Poinsettia, Mona doesn't read that).

Before we look at what we could do, let's look at what we should have done, in time honoured game-show tradition. That at least is much easier. Not have had an industrial revolution, not have allowed the population to grow unchecked into the tens, then hundreds of millions, then billions (and soon tens of billions). Not have cut down the forests of the world to make room for all those miserable privileged people who expect something in return for being born.

What should we do next then? After all, it's highly, almost overwhelmingly likely that most of the damage has been done. $CO_2$, ever the sinister by-product is already at record levels, and stopping now wouldn't stop the rise from record to scary levels by 2100. In a recent article in Science journal, leading biologists stated that governments around the world should protect a third of the oceans and land by 2030, and half by 2050. It won't be enough to stop the $CO_2$ build up and temperature rise, but it would stand a chance of preserving some future for our descendants. At the time of writing that was more than a year ago, plenty of time for a few governments to have shown even a modicum of interest, to have formed a few committees to explore the possibilities of producing a white paper or two. Have they? No, of course they haven't.

But what if I was in charge? If everyone on this planet obeyed supervillain me without question? If whatever costumed vigilantes who could thwart my plans have been safely despatched, and the governments of the world were falling over themselves to do my bidding.

A drastic and immediate population adjustment, that's what. This is why I find contemporary thrillers and near-future science fiction work in literature, TV and film so fascinating, inevitably the writers of said work have had exactly the same thoughts as me and are expressing them through the elegant medium of supervillain fantasy.

Think about it, many of these villains in media (and by proxy, the writers) have seen the writing on the wall and unlike the writers who may go so far as to recycle their whisky bottles in the name of ecosystem preservation, are attempting to do something about it. This leads said villain down a dark and lonely road to a nefarious scheme whereby billions of people will bite the bullet, only to be saved at the last minute by a hero who is invariably wearing a tuxedo for no adequately explored reason.

To make a decent dent, we're talking maybe 90%, not the 50% that accompanied a certain 'snap happy' cosmic villain's recent temporary triumph in the MCU. As ghoulish as it sounds, 50% probably wouldn't be enough to save us. If maybe 6.75 billion of the 7.5 billion people (at the time of writing – by the time you read this it will probably be much more) would shuffle off the mortal coil, then we'd be talking. They'd leave behind a mere 750 million people behind to occupy all that space and focus on a simple pastoral life (preferably planting a lot of trees) without all the ridiculous carbon emissions that accompany life nowadays. The smartphones would need to be put away, the cell towers shut down, the nuclear power stations safely decommissioned.

If that took place, those that remained might well have a chance of having descendants that persist into the 2200s, or even 2300s. The dark logistics of disposing of the bodies of 90 percent of the worlds' population without plagues… plaguing the survivors will be unthinkable. And believe me, I've thought about it.

But that's never going to happen, supervillains don't exist, even if I was some kind of biological warfare genius intent on a way of effecting such an adjustment, I honestly don't care enough.

Here I sit alone at the end of the world, looking at humanity with the curious detachment of Prospero at the end of The Tempest. One of Shakespeare's quotes (actually from Hamlet) that has always stuck with me has peculiar relevance here: 'There are more things in heaven and earth Horatio, than are dreamt of by your philosophy'. I've always admired the other-worldly detachment of a character who can utter something so profound, yet so completely disconnecting.

What I mean in any case is that of all the amazing, beguiling, fascinating facts about our planet, one of them is true above all others, and is possibly the most important one of all. This planet, and all its wonders and resources, was not put here for our benefit. To think otherwise, as a great many of humans do, is a monumental egocentricity – for those of you who take comfort in some form of deity, I refer to the absurdity of the phrase 'god given'. Related to this, another sickening phrase that has macabre echoes through our violent colonial history is 'Manifest Destiny', which in the cold unflinching light of understanding is a cruel joke we played on ourselves.

Better still, I could paraphrase Shakespeare and turn his writing on its head to make even more sense to our current predicament: 'There are less things in our future than are dreamt of by our science fiction writers'. Significantly less unfortunately, and more is the pity, because I love a good science fiction book.

After all, if you're still reading, perhaps you have got some inkling by now of the point of this otherwise irreverent and irrelevant tome I've taken the time and trouble to create for you.

This is not a warning, or some call to take urgent action.

No. This is a eulogy, an epitaph.

The life and times of the human race.

Such as it is, the Tipping Point has well and truly tipped, and not in our favour. In just a few short chapters we have explored the damage as a species we have chosen to have done to our own ecosystem, how the trees have been cut down, the carbonated remains of an entire previous ecosystem dug up and burned for heat and fuel, their remains choking the atmosphere for centuries to come. Ice is melting, sea levels are rising.

If I was a betting man, I'd put the date of the Tipping Point somewhere around 1820, when the human population soared over a billion for the first time in history, never to look back. Industry of every kind sprang up to support this every expanding population, pouring untold pollutants into the ground, into the water and into the atmosphere.

How long until things start to go bad then? Sociologically speaking, we're well on the way down that road to hell. Right-wing populism holds sway in a depressing amount of countries, and for them climate change is either outright disbelieved, or simply not a priority ahead of furtherment of right-wing ideals or personal power.

Environmentally speaking, my humble laypersons guess would be around thirty years before things start to significantly change for the worse. Before the energy shortages become chronic. Before the internet becomes a privilege not a right. Before the majority of society become and remain unemployed as whole industries go bust overnight. Before food runs short. Before water is rationed. Before air starts to taste bad. Thirty years is my guess, and certainly not more than fifty.

That's time enough for any of you with children to help them grow up into mature responsible individuals and assume personal responsibility for their future and their safety. As grossly unfair as it may sound however, I would strongly advise that you coach them against bringing any future lives into the world. While I pour my martini and stare out at a (somewhat but not completely metaphorical) planet on fire, I can think of nothing worse than having to see children go without, to see them go hungry, go thirsty. To see their glorious potential reduced to queueing up with a ration book for vegetables and water (make no mistake there will be no meat on the menu by then – that will be something only the fabulously wealthy can enjoy).

As I mentioned a few chapters ago, I've taken steps to remove myself from the gene pool, I couldn't sire children even if I wanted to, and don't you dare pity me for taking action to prevent a mistake I'd have to live with. Despite this I still like to wear a Nyami Nyami necklace which represents the Zambezi River God, or Snake spirit of fertility. It was a gift from my brother, and I wear it as a reminder that my hypocrisy knows no bounds.

So now you know, barring the last few pages and 'also by the same author' what this book is about. Now I'd like to cover off why this book is about. I certainly didn't write it as an intellectual thought experiment, and certainly not for self-aggrandizement. I didn't write it just to depress you, though I daresay I may have achieved that periodically here, for which I unreservedly offer my apologies.

No, when I said this wasn't a call to action, I wasn't being strictly true. Rule one, everyone lies. There is a call to action here of sorts. I would like to see this book pulling some kindred spirits out of their comfort zones, and to live out their lives doing whatever things they have always wanted to.

As a result of a life coaching exercise my boss put me on, which I found to be an unexpectedly rewarding experience, I discovered quite recently a superpower I didn't know I had. It could be summed up as finding out what someone has always wanted to do but has been putting off and give them a kick up the backside to do it. As a direct result of this, several very interesting sounding books are on the way that otherwise would still be a pipedream of would-be authors. In addition to that several gym regimes have started/restarted, even a new small business is underway through gentle social probing followed by not too gentle goading.

I'd like to offer you the same gift.

The thing about your life is, from your perspective, whether you realise it or not, you are the most important person in the world, and you only have one life to spend. If you're lucky that's something like thirty thousand days with which to do things that interest you, amuse you, or otherwise enrich you in some other way that I'm too twisted to understand.

One of my annoying sayings (if only because I repeat myself so damn often) is 'Time is the only currency that matters'.

It's one of the truest things an old liar like me can ever say, and I'm sure if you think about it, deep down you'll find some artistic project, some classic car you always meant to restore, some orphaned child in Africa who needs a better life. I have no idea what drives you, but I'm willing to bet that way back in the subconscious, in places you don't often speak about, something does.

Pursuing experiences that you find rewarding, that enrich the soul is pretty much the only reason I can find to be alive. This short life is not long enough to discover the secrets of the universe, even were the universe inclined to share them with you, which it isn't. The age old "What's it all about" is a strictly human conceit, you don't see dolphins or cows or whales or wasps having metaphysical crises about why they have found themselves alive at this time in this strange corner of an otherwise unremarkable galaxy.

To quote the lyrics of the magnificent Queen, and the even more sublime David Bowie (mayhap I have mentioned him already?), from their unbelievably poignant 'Under Pressure': "It's the terror of knowing what the world is about". If I could humbly suggest a meaning beyond the purest love that was in their hearts, the lyric "knowing what this world is about" wasn't about something at all. It was about nothing.

In mid-2018, social scientist Mayer Hillman, then at the tender age of 86 is long past the point of giving a crap about anyone's opinions of him. "We're doomed", he told a Guardian journalist with a big smile on his face. I should clarify Mayer was the one smiling, not the journalist who was probably wondering what they had walked into.

Hillman has spent the majority of time since he retired from an extremely influential career guiding government after government in Britain in complex and far-reaching transport infrastructure decision making. Specifically, he's been thinking about runaway climate change. His contention is that the most pessimistic estimates of the impacts of climate change on our ecosystem are far, far too optimistic, and reading his words it's hard to disagree with him.

Interestingly, Hillman goes on to explore a fascinating analogy, that if we're to accept that our civilisation is doomed (as my new best friend publicly states), then it could make humanity behave like an individual who receives a terminal diagnosis from the doctor. Hillman argues that people in this terrible situation don't go off the deep end and go crazy; instead there is a broadly calm and rational attempt to prolong what life they have left. In this way humanity could try and see out the remainder of its tenure as caretaker of this planet with as much dignity as it could muster.

I wholeheartedly subscribe to this recommendation, and I would invite you to consider that assuming you believe Hillman and I, that such a reservation should rather than being not depressing, could be considered very liberating. You know more about the world now I hope than you did before you picked up this book, quite possibly more than many others do.

For the science minded among you we may also be finally witnessing history here in terms of a solution to the Fermi Paradox. Named for Enrico Fermi, an uncharacteristically droll and notorious prankster physicist, who in 1950 uttered the fateful words 'Where are they', during a discussion of UFO sightings and faster than light travel.

The paradox is the contradiction between on one hand, the total lack of evidence supporting alien visitors to this planet, and on the other hand, the age of the universe and assorted credible (and largely statistical) estimates on the amount of life in the galaxy. Alien life it is reasoned, should be prevalent after all these billions of years of time for them to evolve, yet we see no sign of them anywhere. In fact, given the age of the galaxy verses typical evolutionary timescales, life capable of expanding across the galaxy should have emerged a hundred times over by now, yet here we sit by our ham radios, twiddling with our knobs and wondering why nobody is talking to us.

My answer, which to be fair has been stated before by people much more academically accomplished than myself, is that all those evolved species across the galaxy, just like ours, have a tendency to consume their available planetary resources and civilisation burns itself out before becoming interstellar. I'm again reminded of my narrowing tunnel recurring dream, though I doubt my subconscious mind is that sophisticated – I know my conscious mind isn't.

Excuse me for a moment, I need to go and write some more on the unstable weather chapter – you didn't think these things fell together in sequence, did you? I can't speak for anyone else, but that's definitely not how my mind works.

Back. Did you miss me?

Taking a break from all the doom and gloom, the ray of optimism I see in our species future is Elon Musk's Mars colony and as an outside possibility, a Chinese Moon base. Musk has announced plans to have a Mars base up and running by 2028, and while he just loves running his mouth on Twitter, he's also not given to hyperbole. When he makes these statements, he means them. See you on Mars, Elon. Given (conservatively) a few dozen payload deliveries to Mars could happen before we lose the capability for spaceflight, probably in the nationalisation of the space industry and it's dismantling for resources to power homes and feed the hungry, there is time enough.

Time enough for a second chance for the species to attempt to embrace the new while avoiding the mistakes of the old, under the watchful eye of Phobos and Deimos.

If what I've written has struck a chord with you, there's one more thing I'd like to highlight before we wrap up, one thing that feels important for me to say.

This is not your fault. Take that feeling of guilt and stow it somewhere with the rest of your checked luggage. You couldn't have done anything differently, if you're anything like me, you have lived a good life up till this point, and that won't change just because some idiot threw eighty-odd thousand words at Microsoft Word to see what would stick. It's the lie a lot of us hear growing up, the fallacy that there is a Grand Scheme of things – then when we're older we understand the truth of things, it's not grand and there definitely isn't a scheme.

And just because I've made this prediction, this line in the sand, where I stand on one side with my deckchair and trusty martini, it doesn't mean that humanity won't fight till its last breath. And who knows, if it doesn't get too distracted fighting itself, maybe there's a chance, something I haven't seen, some shred of goodness, of ingenuity, of collaboration that I missed. I've been wrong before, and there is an outside chance I will be again after all.

And if nothing else, there is a nobility in going out swinging.

Thanks so much for reading. I'm not sure I'll write anything else after this. But I did say that last time.

In the spirit of going out the way I came in, I'd like to quote someone much smarter than I, the immortal Joss Whedon, who in the final minutes of the script for 'Age of Ultron' had this to say on the subject of humanity being doomed.

A thing isn't beautiful because it lasts.

# ABOUT THE AUTHOR

Tim Biddiscombe is a writer, producer and director from the South coast of England. When not penning diatribes about the human condition he can be found roaming the wilds of Richmond, Surrey.

He believes it is traditional to mention dependents here, so he has a very demanding plant named Mona. He also finds writing in the third person a very perplexing experience, and cannot wait for it to